WATER
The Wonder of Life

WATER

The Wonder of Life

RUTHERFORD PLATT

Line drawings by Stanley Wyatt

Prentice-Hall, Inc., *Englewood Cliffs, New Jersey*

Dedicated to
the late E. Ross Anderson
who loved the power and beauty
of water and wind

Other Books by Rutherford Platt

This Green World
The Woods of Time
Discover American Trees
Pocket Guide to Trees
Wilderness, the Discovery of a Continent of Wonder
The River of Life
1001 Questions Answered About Trees
Worlds of Nature
Secrets of Life
Adventures in the Wilderness (WITH HORACE ALBRIGHT)
The Great American Forest

Preface

Twenty-five hundred years ago, the Father of Greek Philosophy, Thales of Miletus, founded his school of thought on the basic premise: "All things are water." Nothing that the technical advances of the intervening centuries have revealed, nothing that the development of the electron microscope or atomic energy or satellite travel has disclosed has diminished the place and importance of water. In the sap of plants, the bloodstreams of animals, in rainfall on the surface of the land, in rivers flowing to the sea, water represents the great circulation system of our planet. It has covered the earth with evolving life, and its presence, so far as science can determine, makes Earth unique among the planets.

In threading among the complex discoveries that have been made concerning the nature of water, its remarkable properties and the multitudinous life that had its beginning in it and that depends upon it, Rutherford Platt has employed the same special gifts for apt analogy, imaginative comparison and lucid explanation that have made previous books of his highly esteemed in the field of natural-science writing. He has been able to translate the involved and difficult into the seemingly simple and understandable.

I suspect that at no other time in our history would such a book as he has written have been of so great interest and concern to so large a number of people. In this time of poisoned rivers and dying lakes and polluted seas, we have been forced to a fresh appreciation of Earth's priceless heritage of water.

In the spring of the year in which this book is published, I walked along the beach on Padre Island on the Texas coast where the wave-washed sand extends southward for a hundred miles. Once clean, it is now frequently disfigured by masses of crude petroleum stranded by the tides. Thor Heyerdahl reports sailing his papyrus boat through a mid-ocean area of drifting nodules of crude oil so vast it extended for 1400 miles. NATO officials were recently told that 5,000,000 tons of such oil is dumped onto the oceans annually by bilge-pumping from tankers. In the not too distant future no beach in the world will be free from such oil pollution.

And tankers are a comparatively recent innovation. So are the pulp-paper mills that have been pouring their deadly mercury into clear wilderness streams. So are the DDT and other residue pesticides whose poisons drain from the land down watercourses into the ocean. So is the thermal pollution of hot water discharged from atomic generating plants. A scientist concerned with investigating the environmental effects of thermal pollution tells me that it is entirely possible that in a generation's time every sizeable river between the Atlantic and the Pacific will be carrying the heated water of such plants down to the sea.

In the apparently boundless water of the oceans, the disastrous consequences of poisons from the polluted land have already become apparent among floating plankton, among fish far from shore, even among penguins of the Antarctic. During seemingly illimitable stretches of time the seas were undisturbed by any except nature's influences. Now in the space of a few decades, pollution and over-fishing have killed, according to Jacques Yves Cousteau, forty percent of the marine life of the oceans. An unprecedented revolution is disrupting the life of seas and lakes and streams of this water-dependent planet. We are but feeling the first shock waves of disaster.

Anyone who reads this book will lay it down with fresh awareness of the wonder and value of water and with increased alertness for its protection as the Earth's most valuable asset.

Edwin Way Teale

CONTENTS

Preface   ix

1. The Watery Planet   1

2. The Accidental Discovery of H_2O 12

3. Two H's and One O   19

4. The Fabulous Fabric of H_2O 33

5. Water Assembles Life   51

6. The Womb of Life in the Ocean   61

7. Nothing Else Like This in the Solar System   74

8. From Blue-Green Algae to Man's Brain   87

9. The Ocean Endows Life with Sensitivity   98

10. The Two-Animal Animals   111

11. Tape Recordings About Sex in the Ocean   125

12. Love and War Under the Waves   136

13. Fantasmagoria   147

14. Preparations for Going Ashore   166

15. The Call of the Land   183

16. The Call of the Deep   203

17. Twilight on the Delta   221

18. The Great Awakening   234

19. But Thermal Pollution Is Different   245

Index   267

WATER
The Wonder of Life

1

The Watery Planet

Stepping over the rail of the schooner, we walked and ran on the sea ice, eager to stretch our legs. Fresh water aboard ship was low and stale, and sparkling puddles on the ice pans looked tantalizing like the purest spring water. A man flopped on his stomach and cautiously took a taste. He glanced up with an expression of happy surprise, and then quenched his thirst with great gulps. Delicious! I had never heard it mentioned before—that, in some mysterious way, when ocean water freezes it loses its saltiness, becomes purest fresh water.

That little discovery made by this novice polar explorer fifteen years ago may have triggered this book by jogging my curiosity about this common thing we call water. Common it surely is, in oceans, clouds, rivers, lakes, rain, and snow. It pervades us and we dwell within it.

Today's space science is saying that the liquid state is very rare in the universe, that water is the only liquid that forms naturally on the earth's surface, that an H_2O molecule—there are trillions of them in a drop of water—is not simply a particle of water but a precision instrument comparable to a transistor in the way it can organize electronic circuits that awaken energies and transform things.

The powers and works of the water molecule are more fan-

I

tastic than those of Neptune of ancient fable, who could shatter rocks with his trident, and call forth and subdue storms, and shake the shores with it. How astonishingly apropos! What is more, the H_2O trident, with two atoms of hydrogen and one of oxygen, is by far more versatile and performs more fabulous feats than Neptune's trident.

Boiling merrily, dissolving swiftly, clinging to things so as to make them wet, spiraling down a drain, meandering instead of flowing straight, curling and crashing on a beach, transforming into exquisite 6-pointed snowflakes—these are a few of the outward signs of hidden powers of water.

The news about water coming out of today's laboratories has the fascination of outlandish science fiction in the sense that water performs seemingly impossible feats, when you stop to think of it. The supreme "impossible feat" of water was to assemble the wild, disorderly elements and to form a living cell and then sustain the quivering, delicate life stuff in flowing systems while cells evolved and created fantastic botanical gardens and menageries on a small globe whirling through frigid space.

THE WATERY PLANET

Because more than 70 percent of this ball of rock is wrapped in water, the earth might look to an observer in outer space like a big drop of water. The shine of oceans would make the continents merely dark blotches—a tantalizing puzzle like the "canals" in Mars. In contrast to the other planets of our solar system, which are dead dry, earth would sparkle like a diamond as it swings in orbit around the sun.

The observer "out there" would doubtless be excited and incredulous at such an extraordinary sight as a watery planet. On the other hand, we here on earth who are involved with the "big drop" have not regarded it as extraordinary. This is not because of the brilliant charade of H_2O—"playing it cool" while possessing unsuspected cunning—it is because human beings naturally have the same viewpoint toward water as do fish—it is the natural environment. This is also the land creatures' acceptance of air and forests. The media of existence are taken for granted as long as they appear to be inexhaustible. At long last confronted with shortages and pollutions, it is time for us to discover that dissolving, boiling, evaporating, and freezing are not just household words. They are expressions of H_2O's keeping us all alive.

WHERE IS WATER IN THE SOLAR SYSTEM?

How lonely we feel! Especially when spacemen probing out beyond show us photographs of planet earth as a lonely little sphere suspended in a black void. An instinctive yearning to find life (hence water) somewhere else in the cosmos drives us to spend billions to walk on the moon, to prepare a herculean trip to Mars, and to shoot radar signals into the solar system and hopefully detect some incoming ones. So frenetic is this search for company, that even a microscopic fossil impression of a virus on another planet would make big news, or we would even settle for finding sedimentary rocks left by bygone water.

A leading national weekly asks scientists, "If life exists elsewhere in the universe, what do the extraterrestrial beings look like?" "Possibly like humans," says Professor George Wald of Harvard's biology department. "Possibly like plants," says Dr. MacDiarmid of University of Pennsylvania's chemistry department. And so it goes.

Professor Wald explained his reasoning at a meeting of the American Philosophical Society. "All living organisms are almost certainly made of the same four elements—carbon, hydrogen, nitrogen, and oxygen." He points out that all life needs *water*. These specific requirements dictate specific molecules for the same physical functions. Withal, until the 1970's, not the slightest valid evidence has been found to dispel the dismay that life on earth is an unique curiosity, at least in our solar system.

People were betting on the white polar ice caps which fluctuate with the seasons on Mars to supply melt water. They are now known to be frozen carbon dioxide, "dry ice," which by

3

itself would suffocate life. The warm redness of Mars, the planet "most likely to have life," is now attributed to desert sands and dust mixed with iron oxides. There was a cheery hope in 1964 when it was announced that the Mars atmosphere has a water content, even though "it must be about 1/1,000th of that over the driest deserts on earth." Dr. Abelson, editor of *Science*, corrected that to "one millionth" of the water vapor of our desert air. A recent verdict is that it is not H_2O at all but *deuterium*, the "heavy hydrogen" that goes into the making of hydrogen bombs.

Water *ice* has been detected in Saturn's rings—where the temperature is 372° below zero. Venus hidden under heavy clouds was surveyed in October 1967 by spacecraft Mariner 5, which reported that the surface temperature of that bright planet is 800°F, and that the clouds are fine dust resulting from that lethal heat, not ice crystals as had been hoped.

WHERE DID WATER COME FROM?

Since not a drop of water has been detected with certainty elsewhere in the solar system, here is a most fascinating question for the space sciences—where did our oceans come from?

Until recently, it was supposed that the planets had originated as sun stuff in an immense flare pulled out by gravity in a near collision between our sun and another sun. When the stranger sun took off into the black beyond, the fragments of the great flare were left in orbit as planets. This plausible idea ran into trouble when the elements in the sun, deduced by new techniques, were found to be in utterly different proportions than the elements on earth.

What interests us are the two elements of water, hydrogen and oxygen. The sun is mostly hydrogen; indeed hydrogen and helium, the two lightest elements, make up over 99.8 percent of the atoms of the universe. So there is no doubt about there having been plenty of hydrogen for water on earth—or is there? Oxygen stands high in the order of abundance both in the sun and on earth. The rocks of the earth's crust are 50 percent oxygen. So there was plenty of oxygen for oceans and rain on the primordial earth.

Yet the earth, according to its elemental nature, is a little stranger on the doorstep of the sun. It is utterly unique, not only in the possession of water, but in its supplies of the whole list of the elements of life. For instance, plants are 90 percent carbon, hydrogen, oxygen, and nitrogen—BUT to create the beautiful pageants of trees, bushes, flowers, and fruits of our world, the recipe for the remaining 10 percent calls for a long list of other elements, namely, potassium, calcium, sodium, magnesium, iron, aluminum, manganese, sulphur, phosphorus, chlorine—there are traces of some 40 elements in the Plant Kingdom. Mammals are even more elaborate assemblages, with some 45 elements detected in their bodies.*

Now we take a quick stride across the complicated calculations of the space physicists to find that the tally of elements in the rocks and atmosphere of this planet available for living things spells a highly exciting revelation—a truly new and incredible concept—about the origin of the earth, and where the oceans came from.

What the space sciences have come up with has to be theoretical, as no reporter was writing it down in a notebook when it actually happened. However, it is educated speculation. Scientists can duplicate in laboratories what is known about the elements and temperatures on the sun and in cosmic space, to observe how atoms behave. In current laboratory language, they made a "model" of how the earth began; when tested again and again, the experiment gives the same answer.

Instead of being a splash of fire from the sun, the earth (and the sun and our other planets) came from a whirling cloud of cosmic dust which condensed in frigid space. Such clouds of electrified particles are not imaginary, they are well known and have been reported by our space vehicles. They are over 99 percent hydrogen and helium like the sun, with important dashes of oxygen, carbon, nitrogen, *and*, in the course of cosmic

---

* In chemical shorthand they are: H, O, C, N, S, P, K, Na, Cl, Ca, Mg, Fe, Mn, Zn, Cu, Al, Co, I, F, Br, Li, Rb, Cs, Sr, Ba, Hg, Cd, B, Ga, Sc, Y, La, Ce, Nd, Gd, Dy, Ti, Ge, Mo, Sn, Pb, As, U, Se, Ni.

5

time the other elements are manufactured by atomic energy inside the cloud of cosmic dust.

In the process of becoming our solar system, the whirling cloud was a chaos of radioactive and magnetic energies in which internal temperatures rose above 3,600°F. This started *thermonuclear reactions*, induced by the violent collision of the nuclei of the hydrogen and other atoms. This is what creates the radiant energy of the sun—but the part of the cosmic dust cloud that became planet earth was a paltry incident of outer space compared to the massive magnificence of the sun.

The earth's cosmic dust cloud turned into a terrific turbulence in which myriads of ionic (electronically active) particles collided with myriads of atoms, and hydrogen changed to helium, which in turn was transmuted to nitrogen, and chain reactions through eons forged many different elements. When this ferocious dust cloud cooled down—its crust hardened—it found itself in orbit with the sun and a string of other planets.

AND THEN THERE IS WATER!

According to the laws of chemistry and physics, which are the same everywhere in the universe, there should be no water on the surface of our planet, certainly not oceans. For one thing, fluid water can occur only in an extremely narrow temperature range—between 32°F (the freezing point) and 212°F (the boiling point). This water-permitting temperature exists only for a rare fleeting instant in a universe where temperatures range between the deep freeze of outer space and the thermonuclear heat of stars. Look at the glaring desert on the moon, where temperatures rise to 215°F by day and drop to 200° *below zero* at night.

Apart from temperature, however, there should be no water on earth because it takes hydrogen to make the H_2O molecule. Hydrogen, the lightest of all the elements, should fly away into space. Small earth does not have strong enough gravity to hold on to free hydrogen. As the cosmic dust cloud condensed, most of its hydrogen did, in fact, take off into space. This was lucky for us. If the forming globe had been big enough to hold on to its 99 percent hydrogen it might have become a star.

The sequence of events that saved enough hydrogen for oceans and made the planet comfortably cool is a fine example of how natural happenings are hardly distinguishable from miracles.

The theory is derived from the physical evidence of a globe that has a very hot iron core, insulated by a rock crust in which all the elements are locked up.

When the violent magnetic fields of the cosmic dust cloud quieted down (in the course of infinite time), its substance condensed and transformed to rock crystals. This captured much hydrogen before it was lost to space. The hydrogen imprisoned in the rocks is bound with oxygen in many combinations containing carbon, silicon, phosphorus, sulphur—and with minerals such as iron, magnesium, potassium, sodium.

In addition to this hydrogen secured in rocks, the newborn planet saved much hydrogen in the lighter gases of its atmosphere, where it is combined with carbon and nitrogen as methane gas and as ammonia. Scientists believe that earth's first atmosphere consisted of these two poisonous gases.

In this brief glance you catch sight of the elements of life stuff in a future era—along with a good deal of hydrogen.

We behold planet earth being stocked with atoms and molecules that will preserve all the elements life will need, in the way a grocer packs a bag of groceries with the heavier things at the bottom and the lighter things above. The core is iron and nickel; the lighter rocks, namely the granite of continents, float high like crustal foam, are half oxygen, and have silicon, aluminum, magnesium—limestone is even lighter, with calcium, carbon, and oxygen. Nitrogen and carbon, oxygen and hydrogen, are floating around in the gases of earth's original atmosphere.* Life is going to live on the surface and find its most important elements on hand there.

---

* The earth's original atmosphere was almost wholly lacking in free oxygen. It consisted of such simple gases as methane, water vapor, ammonia, and hydrogen; and later, carbon monoxide and dioxide, and nitrogen.

And then there is fluid water—an unimaginable chance occurrence. We might call it a stroke of luck that when the earth cooled enough to have a crust and become a satellite in orbit it was *exactly the right size* for its gravity to hold on to light fleeting hydrogen atoms when two of them were attached to an oxygen atom.

Along with this, there was another towering coincidence. The earth came into orbit at *exactly the right distance* from the sun, to enjoy an average temperature within the very narrow range between freezing and boiling. Life can exist only in this precise place, this exact distance from the sun.

Other planets didn't fare so well. Mercury, orbiting nearest to the sun, has an average temperature on its sunny side of 700°F—far, far hotter than the boiling point. Water wouldn't last there a second. Venus, on an orbit closer to the sun next to the earth, has been explored by a space vehicle that reports temperatures high enough to melt lead. Mars, our nearest neighbor, is just a little farther from the sun than we are. That little difference gives Mars an average temperature of 41° below zero. Jupiter, just beyond Mars, is 150° below zero.

Saturn has an atmosphere made of frozen ammonia, and its crust is a layer of "ice" calculated to be 14,000 feet thick. The surface of Uranus is 305° below zero. That of Neptune, a bit further out, is 325° below zero. Pluto wasn't even discovered until 1931. Forty times farther from the sun than earth is, this daughter is truly in the grip of frigid outer space.

In short, only at a critical distance from the sun can there be water—and even there water will not turn up and persist unless the space vehicle is just the right size (has strong enough gravity pull) to grasp plenty of hydrogen and oxygen before they fly off into space.

As we have seen, planet earth was a particular combination of cosmic forces. As an atom smasher it created a diversity of elements, and these made chemical combinations of molecules which crystallized as the rock crust. In so doing, lots of oxygen and hydrogen was secured, in or near the surface or in the atmosphere, together with a good choice of other elements for living cells.

8

Yet, such a phenomenon as life could never occur on a planet where hydrogen and oxygen—as well as carbon, nitrogen, phosphorus, etc.—are locked up in rocks and in methane and ammonia gases. For that fantastic outcome there must be flowing, globe-circling *oceans*.

THE OCEANS GATHER

Soon after the earth cooled enough to have a crust, water must have gathered in scattered, shady hollows of a basalt desert riven with rivers of fiery lava. The water sizzled on hot rocks, and steam carried off heat into space, where water vapor mingled with the methane and ammonia in earth's first atmosphere. The cooling machine was starting to operate.

If we pause a moment to look through millions of years into the future, we see algae, the green pastures of the oceans, pouring free oxygen atoms into the atmosphere, thus changing the poisonous first atmosphere into the breathable air we know.*

Through millennia the earth's face, pockmarked with volcanoes, cleared; the volcanoes subsided, the cooler areas constantly spread, puddles became lakes, rivers began to run and waterfalls to roar. Water was taking over the face of the planet, and the ocean basins were slowly filling.

Moderation of temperatures proceeded as the oceans grew and grew. During that era water gushed in the steam of volcanoes, and it was vomited out of deep fissures in the rocks which tapped subterranean reservoirs. At the same time, heavy clouds formed in the cooler atmosphere, and during millennia the globe was wrapped in a blanket of black clouds, whose incessant cloudbursts returned water to the rising water level in the ocean basins.

What an awesome prologue for the miracle of water and life!

---

* Before the early algae went to work, some oxygen was added to the early atmosphere by the breakdown of water vapor by solar radiance. The resulting hydrogen escaped from earth's atmosphere; the oxygen remained to combine with carbon into carbon dioxide, promoting the first photosynthesis.

The rare apparition of oceans on a small sterile planet, in the darkness below the black clouds, to the accompaniment of thunder and lightning. This era was the "pre-geologic phase" of the planet. It is calculated to have lasted about 1.5 billion years, during which the oceans spelled the success of the astounding conversion of a hot, deadly satellite into a space vehicle for living things.

In a sense the evolution of life began with this prologue—to be precise it began with the evolution of the composition of ocean water. Sterile fresh waters delivered from volcanoes and gushing from the depths of the rocks and by cloudbursts had to be ripened with organic elements dissolved from the rocks and washed out of the atmosphere, to beget life in the first place. To become the womb of evolving organisms, ocean water had to reach an exquisite state of saltiness in which the raw elements created in that whirl of cosmic dust would not be frozen in ice or remain forever inert in rocks, but would be released and tossed together in the great ocean mixing bowl where they would collide, unite, break apart, collide again and again, in transitory molecules (some became magnificent as protein molecules)—effecting millions of combinations of themselves every second. Life-sustaining sea water is a chemical quality far more complicated and dynamic than a mere solution of sodium and chlorine, which gives ocean water that table-salt taste.

And then water exercised its unique atomic powers to assemble, organize, and sustain the right molecules in the right places for a living cell. It does this with water *systems*, in which there is a constant interplay of the elements of life.

It may be a global system, such as the great water cycle between ocean, ground, and air. It may be an ocean river system, such as the Gulf Stream, the Japan Current, the Humboldt Current, the amazing immensely powerful Agulhas Current in the Indian Ocean. It may be a river system on a continent, like that of the Upper Amazon, where mysteries of life in the "steaming stillness of an orchid-scented glade" are still being discovered. It may be a warm bloodstream system in an animal body. Or—it may be a microscopic system inside a living cell that is a million times more intricate and precise than the finest watch.

This is the exciting story of water and life. Let us look into it.

2

The Accidental Discovery of H₂O

Water was water, element of the ancients, right up to modern history. In 1732 the chief authority on chemistry (Boerhaave's *Elementa Chemiae*) gave this scientific description of water: "A very fluid, scentless, tasteless, transparent, colourless liquor, which turns to ice with a certain degree of cold." Amidst all the excitement about the chemical elements at that time—sulfur, phosphorus, mercury, gold, etc., and frenzied chemical activities of glassmaking, metallurgy, cookery, wines, painting, enameling, dyes—water was just there, an indivisible entity beyond the reach of the chemical arts.

Hydrogen (not yet called "hydrogen") and oxygen * were recognized as explosive gases, but their relation to water was unsuspected. To amuse his guests, a scientist filled soap bubbles with hydrogen and with boyish enthusiasm watched them rise into the air. Joseph Black, a professor of chemistry at Edinburgh University, invited friends to dinner and then said he had something to show them. He liberated the hydrogen-filled bladder of a calf, which promptly hit the ceiling. The company was sure it was attached to a black thread pulled up through a tiny hole by an assistant in the room above.

The next step in the "scientific analysis" of hydrogen came

---

* Oxygen was discovered by John Priestley in 1774.

when the first hydrogen-filled balloon was released from the Champs de Mars in Paris in 1783. As our television announcers would say, "The regular program is interrupted for this bulletin just received. . . ." The balloon rose buoyantly to a considerable height as people gasped with astonishment to see the big sphere (13 feet in diameter) defy gravity. An hour later it collapsed and fell like a writhing rag into a field 13 miles away, where peasants were "filled with terror at so foul and monstrous a bird."

A bitter argument ensued about sending a man aloft. King Louis XVI declared that, if humans were to take balloon trips, criminals should be selected, as their lives were less important to the state. French scientists were indignant "that vile criminals should have the glory of being the first men to rise in the air." Straightway the loudest protester, a chemist named Pilatre de Rozier, actually soared a few hundred feet up into the air dangling among ropes under a bag filled with hydrogen gas as the crowd roared.

Only a few weeks later—with the acclaim for history's first aeronaut resounding in his thoughts—de Rozier was again aroused to make a dramatic protest to a report from across the channel, which he, as a chemist of France, knew was preposterous. Henry Cavendish, a rival English chemist, announced in effect that water is not water! It is not the fluid which flows and splashes in the everyday world. It is not the element which evaporates as water vapor, condenses to form clouds, and is reborn in rain. Cavendish affirmed that water is "inflammable air"—at least it transforms into that in a flash. Cavendish had found this out in an unexpected way when he wasn't studying the nature of water at all.

Electricity was the scientific sensation of the day. Benjamin Franklin had flown his kite and discovered that lightning is a mighty electric spark, and this had led to the invention of lightning rods. Scientists were contriving all sorts of experiments relating to electricity just to see what would happen. They were magnetizing amber, glass, and sealing wax by rubbing them with wool. Galvani, an Italian professor of anatomy, made the muscles of a freshly killed frog twitch by bringing its moist body in contact with touching metals of iron and copper. That led to the invention of the galvanic battery, which generates a current of electricity at a steady voltage. In contrast to such exciting experiments, Cavendish's was incidental, just to satisfy his curiosity.

He wanted to find out what happens to an electric current that passes through water. How does the water affect its voltage? It was a simple matter to fill a glass tube with water, seal it to make it airtight, attach an electrode at each end, and shoot an electric current of a certain voltage through it. He could try tubes of different dimensions and make a chart of what happens to the electricity. The water was of no interest in this experiment, that was only the conductor—nothing would happen to that.

So Cavendish sat on a chair before his simple setup, switched on the current from a "modern" galvanic battery—and saw the water in the tube vanish before his eyes. He was astounded. Had the water somehow evaporated in a split second? Had it turned into air? But that was impossible—unless the tube was not quite airtight. It *must* have a tiny leak.

He repeated the experiment again and again. Each time the water in the tube vanished. He tested the apparatus—there was no defect. So thereupon, Cavendish analyzed the invisible contents of the tube. As the outstanding English chemist of his time he could do this with skill and precision. He discovered that the gas in the tube was a compound of hydrogen and oxygen in the proportion of two to one. He quickly discovered that this curious gas, a product of water which is a quencher of flames, was highly combustible, that it stimulates fire. He named it "inflammable air" in his paper that was published by the Royal Society, January 15, 1784.

As we shall see in a later chapter, the water molecule is a tight, almost indestructible unity of hydrogen and oxygen. It can be sundered only by radiations known to twentieth century science, and by electrolysis—shooting a sharp electric current through the water.

It so happens that at the same time Cavendish was unwittingly making this momentous discovery, James Watt, the engineer who is famous for his inventions of pumps—and whose name is given to the volt-ampere electric power unit—was shooting electricity through water and being similarly astonished to see the water disappear. Watt's paper on this subject was submitted to the Royal Society on April 29, 1784. He was surprised and "sore irked" that Cavendish had forestalled him. However, he was not as irked, it seems, as Pilatre de Rozier when the news about Cavendish's inflammable air got across the channel.

The world's pioneer aeronaut knew a thing or two about gases. Had he not dazzled the public with his hydrogen balloon? The papers of Cavendish and Watt published by the Royal Society made "inflammable gas" sound like hydrogen gas. This is patently absurd. Of course, when water is boiled it transforms into invisible water vapor—but that is not explosive, and it reverts to liquid when cooled. To assert that pure, cool water sealed in a glass tube turns into an inflammable gas like hydrogen at the touch of electricity is preposterous. This de Rozier proceeded to expose with showmanship which, like his balloon, was worthy of a TV presentation.

De Rozier's setup duplicated that of Cavendish in every detail except that he attached a valve to the glass tube with an arrangement to inhale the so-called inflammable air. Then, before a panel of prominent citizens as witnesses, de Rozier switched on the electric current, filled full his lungs with the gas in the tube, and held a lighted taper at his lips as he exhaled —"There was a terrific explosion and de Rozier thought that all his teeth had been blown out." That is all there is in the report about the event which I have read. So the incredible fact that water is a liquid state of explosive gases was confirmed. The history of chemistry gives equal credit to Cavendish and Watt as the discoverers.

After de Rozier's standing as a chemist blew up, a towering French scientist, Antoine Lavoisier, entered the hassle over water, which had now become an international issue. Lavoisier enjoyed a reputation on both sides of the channel for his work in mathematics, botany, and chemistry. He had been a member of the prestigious Académie Française for sixteen years at the

15

time of de Rozier's mortification; he had the authority to revise scientific nomenclature.

After electrolyzing water in glass tubes in his private laboratory, Lavoisier was ready with an unequivocal answer to the question, "What is Cavendish's inflammable air? Is it perhaps a peculiar kind of water vapor? *"Mais non, messieurs, voici hydrogen!"*

Lavoisier explained that hydrogen was the perfect name for this element of water—*hudor*, the Greek word for water, is combined with an abbreviation of the Latin word *genero*, to generate, to beget. In short, hydrogen is *water producer*.

There was no disagreement among those electrolyzing water in glass tubes that there is twice as much hydrogen as oxygen in the "inflammable air." So, on the authority of Lavoisier, water was dubbed H_2O. Ten years later this distinguished member of the Académie Française fell victim to the French Revolution and was guillotined, and our telling idiom H_2O became his epitaph!

As a milestone in the progress of probing nature's secrets, the splitting of a molecule of water into two H's and one O is comparable to the splitting of the atom in our time. Prior to this, scientists had no clues to the composition of water—see Boerhaave's description on page 12—and H_2O led to the recognition of the elements and modern chemistry.

Then, as now, people were excited by visions of science in a changing world. Newton had watched the apple fall from a tree in his garden at Woolsthorpe and had given the world the law of gravitation and laid the foundations of astronomy. Christopher Wren's St. Paul's Cathedral had arisen on the ashes of London, and a new style of architecture was changing the face of the cities. Moreover, scientists were alarmed about their discoveries. In the standard chemistry textbook of the time it was written: "God grant that mortal men may not be so ingenious at their art as to pervert profitable science to horrible uses . . . in whatsoever age they live; there being still more surprising things hidden in the secret powers of nature." *

* John Read, *Humour and Humanism in Chemistry* (London: G. Bell & Sons, Ltd., 1947).

H$_2$O was the breakthrough that led to science's coming to grips with the atomic nature of matter. But consider how the dramatic accident left scientists reeling. They had proved an incredible fact—that hydrogen and oxygen, ethereal, explosive gases, turn into water when they combine in the proportion of two to one. But *why?* What is the force that makes them stick together? Is there here an arcane law of nature? The time was ripe for a man who could answer these exhilarating questions.

John Dalton was 18 years old when Cavendish saw the water in his glass tube vanish and de Rozier blew his teeth out. He was teaching primary grade youngsters in a Quaker school "English, Latin, Greek, French, writing, arithmetic, accounting, and mathematics, for ten shillings and sixpence a quarter." This fits the thesis that really creative intellects have a broad spectrum of cosmopolitan interests, "Really gifted creative individuals are attracted by almost any problem in almost any field. . . . They are often open-minded to the point of gullibility in accepting bizarre ideas; at least they will toy around with such thoughts seriously before discarding them." * Pupils at Dalton's school remembered him in later years as often lost in thought —"always jotting down ideas on scraps of paper and throwing them down absent-mindedly."

I think it is apropos that Coleridge and Wordsworth, who were making poetic discoveries of nature, were contemporaries of young John Dalton, whose identification with nature was deeper in some ways than that of the "lake poets," according to his biographer.† Furthermore, in the same year that Dalton was pondering H$_2$O, "trying to understand why the forms of nature are as they are," he heard the mystery and magic of Beethoven's *Moonlight Sonata* and *Choral Fantasy*.

When Dalton proposed a system of symbols for each element, he was considered a philosopher, not a chemist. A contemporary chemist commented "Mr. Dalton is too much of an atomic philosopher." This would not have displeased him. His first account

---

* Lewis E. Walkup, *Individual Creativity in Research*, paper published by Battelle Memorial Institute (Columbus, Ohio, 1958).
† Frank Greenaway, *John Dalton and the Atom* (London: William Heinemann, Ltd., 1966).

of the atomic theory was published in 1808 under the title, *A New System of Chemical Philosophy*. Much of what Dalton said is now revised, and many niggling objections have been raised. Yet Dalton—triggered by H_2O—set modern chemistry on its feet because *he had the right idea*.

Until recently the water molecule was visualized merely as two small atoms of hydrogen clinging to a much bigger atom of oxygen. The pyramid structure of H_2O's and their electronic energies were unknown until our atomic age. We shall return to that fascinating subject in a later chapter.

3

Two H's and One O

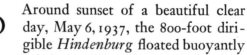

O Around sunset of a beautiful clear day, May 6, 1937, the 800-foot dirigible *Hindenburg* floated buoyantly up to its mooring mast at Lakehurst, New Jersey. It was the triumphant completion of the maiden voyage with 97 people aboard, hailed as beginning a new era of transatlantic travel.

So fine was the weather that the captain had steered a course to display his silver ship in the blue sky to cheering throngs along the shores of Long Island Sound, and had twice circled New York City. Now at the Lakehurst mooring mast a thousand spectators shouted, the zeppelin dipped its colors, smiling people waved from the windows of the luxurious cabin, newsreel cameras clicked merrily—as in a split second the gleaming marvel up there *vanished in a fiery inferno.* The duralumin gondola melted and writhed like a ghastly serpent as it plunged to the ground, the mercury in the thermometer of the Naval Air Station shot upward, and an immense cloud of "white smoke" mounted into the clear air above the holocaust.

A radio announcer who was an eyewitness, trying to describe the disaster later, mumbled something about "spooky white smoke" and then broke down and wept. The "white smoke" billowing out of the bag was mentioned by all the news services—normal black smoke came later from burning fuel and debris on the ground.

The "white smoke" was not smoke at all—it was like steam from a kettle spout, or like a cloud in the sky that forms from the condensation of water vapor touched by cool air. But how could such a great cloud of steam erupt from that magnificent silver bag in which there was no water to boil in the flames?

The surprising answer emerged at the investigation when men from the Naval Air Station who had been assigned to grasp the landing lines casually mentioned that they had heard one of the engines backfiring as the airship nosed up to the mast. According to standard practice the captain had ordered hydrogen to be valved to steady the airship for mooring. Ordinarily the hydrogen gas escaping through the valve would have dispersed invisibly in air currents—but expert consultants deemed that this time the free hydrogen must have been ignited by a spark from the backfiring engine. This caused a flash of fire that tore a hole in the bag, which was instantly rent, releasing 7 million cubic feet of molecular hydrogen into the vast reservoir of free oxygen in the air. In short, the *Hindenburg* calamity staged a spectacular exhibit of the combustion of hydrogen by oxygen —the natural process that created the first water on the primitive globe.

When the crust of the newborn planet hardened, as we have seen, a big amount of molecular hydrogen was imprisoned in the rocks. When quantities were set free by volcanic action, the pure hydrogen gushed into the atmosphere, where this highly combustible gas collided with oxygen freed from rocks rent and shattered by earthquakes. Unattached H's and O's always snap together with electrostatic attraction because they brandish opposite electric charges—hydrogen beckons with a plus sign, oxygen with a minus sign. Chemists say that when they unite, the hydrogen is *oxidized*, the technical way to say that hydrogen is "burned" by oxygen.

Burning in the everyday sense leaves ashes. When coal or wood is burned it leaves carbon ashes—other elements disappear

20

in the air as gases. When iron is oxidized it is burned in a chemical sense although it does not spring into flame—the ash is red rust. When hydrogen is oxidized, the "ash" that is produced is WATER!

This singular phenomenon occurred on earth after the crust hardened and moderate temperatures prevailed over large areas. Thereafter, the oxidizing of hydrogen proceeded through eons, and the ocean basins of the juvenile planet filled with water.

When this happened in the case of the *Hindenburg,* the heat of the spark quickened a chain reaction, a phenomenon on a tiny scale similar to that of an atom bomb explosion. The product of this was water vapor, which instantly boiled from the flames of its own making and condensed in the cool air to form a cloud. If the man-made cloud had occurred at Lakehurst on a moist, overcast day instead of on a beautiful clear day with low relative humidity, the stunned spectators might have been sprinkled by a quick shower.

You can see the process that blessed our planet with water still occurring wherever volcanic activity is releasing hydrogen gas from rocks deep in the crust. The instant the elemental hydrogen touches free oxygen in air, the two elements dart together by electrostatic magnetism.

Since loose oxygen atoms in the lower atmosphere (the oxygen we breathe) always cling together in two's (O_2), four hydrogens are involved in creating each brand new water molecule. According to natural law, water is composed of twice as many hydrogens as oxygens. But even H_4O_2 is but a chemical concept that has never been detected. In the collision between H's rushing up the flue from the deep caverns of the crust, and the O's swirling down in air, countless more atoms of hydrogen and oxygen join the water-making frenzy.

This new water is visible as "steam" when the water vapor is cooled in the air. It is the eerie plumes that hover over Mt. Etna and Vesuvius. It is the white smoke in the Great Fire Pit of rumbling Halemaumau in Hawaii. Most spectacular is the vision of water forming in Alaska's Valley of Ten Thousand

Smokes, where steam is still rising from thousands of fumaroles left when an incredible explosion in June 1912 flung out 7 cubic miles of rock. The valley, 4 miles wide and 15 miles long, holds no living thing, and the water is born with sounds of elemental energies, of hissing and puffing in the vents and an occasional boom from deep below.

However, this new water created in isolated spots is far less than a drop in the bucket. It would take a nightmare to conjure up a concept of the towering flames and titanic explosions that enveloped the whole newborn planet when the fresh crust was heaving and cracking and finally stiffening with vast depressions for ocean basins which the global tumult of hydrogen and oxygen gases would fill with water in a *billion* years or so.

Moreover, what tiny amount of water may have been added to earth's reservoirs in this way in geologic time is negligible. By geologic time, the water supply of our planet was virtually all formed. It will never increase or decrease appreciably as it revolves through the great water cycle, perpetually passing between its three states—visible water, invisible water vapor, and visible droplets in clouds. In addition, a hefty portion of the planetary water is temporarily locked up in a fourth state, glaciers and polar ice—but at the touch of light these melt away and return their water to the cycle.

So it happens that the water supply on this lonely little sphere whirling in black space is fixed and constant. About the same amount of precipitation occurs each year. It is only the geographic distribution of rain and snow that varies according to the caprice of winds and the pranks of the fluid elements of the ocean and atmosphere. In the fall and winter of 1969–1970, incessant bad weather delivered excessive rain and snow in eastern United States and Europe, while the Australian outback was scorched and ranch animals perished—and Alaska faced grim forest fires in the summer of 1970 from drastic shortage of rain and snow.

On an overall average, precipitation is occurring on only 3 percent of the earth's surface at any given time, while clouds constantly cover 50 percent of the earth's surface. In short, non-precipitating clouds are common, and rain clouds are the exception.*

This stability of our world's supply of water relies on the theory that hydrogen is such a light element it would fly off into space when set free from the crust, were it not captured and held on this planet by combining with oxygen molecules of H_2O. This combined molecule is nine times heavier than an elemental hydrogen molecule; thus, in water vapor, gravity can hold the hydrogen on earth. Fine instruments of the space age detect loss of hydrogens from the upper stratosphere when water molecules are split by cosmic rays, separating the H's and the O's. When this happens in rarefied air at the top side of the atmosphere, about 75 miles above the earth's surface, H's that do not find O's to join have lost their anchors and zip off into outer space.

It is impossible to estimate how much hydrogen escapes from the earth's gravity in this way. Tree rings of fossil wood indicate that average precipitation has been constant for epochs. Experts suppose that whatever water vapor is lost is offset by new H_2O's in the gases of volcanoes still performing the ritual that created the original water puddles of the planet.

Water created a couple of billion years ago is still around! The H_2O's have never ceased to exist. At any given time they may be in the vaults of the rocks, in the black abyss of the ocean, swirling in clouds, or in the transparent vapor of the atmosphere. You are an artifact of H_2O's. Your fingernail may have calcium that was dissolved from the shell of a prehistoric oyster. The iron that makes your blood red may have spent millions of years in iron ore before water dissolved it out of the rock. The water in your coffee pot may have existed for ten thousand years in the glacier that covers Greenland until sunlight set it free and it joined the clouds.

Water is a unique chemical compound. It is the only natural fluid that forms in the moderate temperatures prevalent on the surface of our planet, between $32°F$ and $212°F$. When mole-

---

* This calculation comes from a telling book, Reginald C. Sutcliffe, *Weather and Climate* (London: Weidenfeld & Nicolson, 1966).

cules are locked together with their electricity satisfied by equal plus and minus signs, they are immobilized. We see them thus as crystals in rocks. When atoms are heated up, they vibrate so furiously they cannot make durable combinations. The solid substance disappears as gas. The in-between state is *fluid* —in which molecules quickly make and break connections.

Only chemists who can achieve extreme temperatures and pressures in laboratories ever see elements in the rare state of fluidity.* But when two hydrogens and one oxygen get together, the one fluid that forms naturally in moderate temperatures on the earth's surface appears for everybody to see!

THE INVISIBLE SORCERER

Nobody has ever seen an H_2O molecule. It cannot be pictured as a physical thing except with theoretical diagrams and equations. A water molecule with two atoms of hydrogen clinging to one atom of oxygen is so minute that it is far beyond the range even of the electron microscope which can magnify an image "one thousand times ten thousand." Not only is it too tiny to glimpse, but also its individual existence is so brief that it can only be described in poetic terms, such as "the infinite part of an instant." Nevertheless, thanks to the exquisite techniques of present-day atomic sciences, the actual shaping up of an H_2O can be detected and its magical behavior and powers explained.

Magical but not supernatural. H_2O strictly obeys the laws of chemistry and physics. Consider the way water promotes chemical reactions. With its dissolving power, water is the chief agent for making solutions in the laboratory, and it takes part in centrifuging and precipitating.

Moreover, water is a versatile exponent of the laws of physics. It is like a star player of a football team in kinetics—the dynamics of things in motion that involves gravity, inertia, and collision. This entails concepts of wave mechanics, of radiant

---

* Mercury is an exception.

light and fluids, and of thermodynamics.* Physicists rely on water at every turn in their adventures with complicated matters like kinetics and thermodynamics—and water always works without too great a tax on their manipulatory abilities.

The lawfulness of water enhances its wonder. It confronts us with an essential fact beyond human imagination or scientific insights to explain. It possesses order and creates order. Is not water something on the level of divine creation? The laws which have summed up before our eyes electronic realities such as television, conversation with men walking on the moon, and orbiting satellites that transmit messages, should give pause to anybody who puts all reliance upon the sensory and external.

A CLOSE-UP VIEW OF H_2O

Scientists can plot the atoms in a molecule by shooting X rays through it and seeing how they are deflected; they can photograph the paths of ions in a "bubble chamber," which the inventor says he designed by studying the bubbles in a glass of beer; they use high-speed computers, transistors, and devices that one physicist terms "more witchcraft than science."

So, now behold our magician H_2O and see what endows it with its powers and creativeness. It is a *thing*, a physical substance, although no eye has ever seen it and it never stays put in one spot. Before we examine the combination of two hydrogen atoms and one oxygen atom, consider each of these elements by itself.

Hydrogen constitutes some 90 percent of all the matter of the universe. All other elements are comparatively insignificant on a cosmic scale. However, hydrogen is so light that by weight

---

* The two famous laws of thermodynamics are: (1) The conservation of energy—mechanical work is transformed into heat, or vice versa, and the amount of work is equal to the heat generated. (2) Heat always flows from a warmer body to a cooler body. Heat never, by itself, flows "uphill." The result is *entropy*—the degree of loss or unavailability of heat energy. For example hot water, as in a steam engine, increases its entropy, loses heat energy, in proportion to the work it does. Spent steam will not drive an engine until it is heated again.

it is well down on the list of elements in the earth's crust—of about the same abundance as phosphorus. Hydrogen, the substance of outer space, was stored in the oceans of our planet by combining with oxygen (as noted on page 7), which made two atoms of hydrogen nine times heavier than before this combination.

Due to hydrogen's urge to rise, lower levels of the atmosphere contain practically no free hydrogen, about one-fifth of 1 percent, while high levels of the stratosphere are 99.5 percent hydrogen. It zips around, lively with such strong magnetic attraction, that it unites with the atoms of many elements. This activity creates some curious effects when, by joining much heavier elements, it changes them into gases by its giddy buoyancy.*

Warmth is another attribute of elemental hydrogen. Hydrogen makes the hottest flames. Iron and steel are welded with a hydrogen-oxygen torch. Hydrogen joining oxygen invariably emits a puff of heat. Moreover, active hydrogen, which chemists call ionic hydrogen, is the element that gives acids their bite and sting.

Oxygen by itself is likewise exciting. It is the second most abundant element of the universe. Oxygen is the element of combustion. Our spectroscopes detect oxygen in sun flares and this is the element of the atmospheres of stars and comets. It is the eager joiner and stirrer-up of elements. That is to say, oxygen brings movement—chemical activity—into even the Mineral Kingdom of elements.

A particular service which atomic oxygen renders to life on earth is a flock of O atoms that forms a layer called the *ozone* (O_3) *zone* about 12 miles above our heads. This blanket of O's efficiently traps short-wave radiation consisting of ultraviolet and deadlier alpha, beta, gamma rays that shoot through space and reach our atmosphere without any loss of power and speed.

---

* For example: natural gas (methane) results when 4 H's hitch to a carbon atom. Will-o'-the-wisp (phosphene) is 3 H's with a phosphorus atom. Rotten-egg odor (sulfide of hydrogen) is a 2 H's with a sulfur atom.

If they reached the biosphere where life exists, they would shatter the molecules of life and make it impossible for them to form. Below the ozone zone, the diluted oxygen (O is 20 percent of the air we breathe) supplements the splendid protective shield of ozone, so that only a small amount of ultraviolet radiation gets through. (Enough for a sunburn on a hot summer day.)

Thus it is that the globe is enwrapped by an atmosphere of air, 12 miles deep between the ground and the ozone zone, in which water molecules remain intact in water vapor, clouds, snowflakes, and raindrops.

TWO H'S AND ONE O

As chemical formulas go, H_2O is disarmingly simple, involving only two very common elements with a mere one-two-three atom count. This should be an insignificant molecule indeed, surely one that should cease to exist the instant after it formed, not comparable to the molecules that make firm combinations of elements which compose the rocks and gases of suns and planets.

But an incredible thing about it is that after two hydrogen atoms and one oxygen snap together, an H_2O is practically *indestructible.*

H_2O's can only be destroyed by having their H's and O's torn apart by cosmic rays, or by electrolysis (passing an electric current through them), or by heating up to 5,400°F. (Such a temperature occurs naturally on our planet only below 600 miles under the surface.) As a physical substance, water seems to be "destroyed" when it vanishes by drying up. But evaporation does not do away with a single H_2O. Its molecules are intact and ready to condense and restore the same volume of fluid at any moment, after any length of time.

So what is unusual about H_2O that gives it such wonderful properties and makes it so astonishingly able? It is a remarkable, vigorous interplay of electrostatics between the two hydrogens and the one oxygen which transforms the two elemental gases into the sparkling fluid that has rolled down the corridors of time on its unresting course.

Chemistry is based on the electrostatics *—the science of the energies exerted between atoms and molecules forever trying to equalize their static electricity. The nucleus of every atom has positive charges of electricity in particles called protons. The surface of an atom (called its "shell") consists of negative charges of electricity, called electrons, that orbit the nucleus. So atoms act like little magnets. They attract each other, and join, when they reach out with opposite signs (plus and minus charges). When they reach out with same signs, both plus or both minus, they repel each other, and each zips off separately, flies around erratically, until it collides with another atom, molecule, or ion (a loose particle with an electric charge) that reaches out with an opposite charge, whereupon these unite.

This magnetic relationship between atoms and molecules creates all matter, with temperatures, pressures, and radiant energies calling the tune.

Let us now assemble an H_2O out of hydrogen and oxygen atoms in mid-air!

Free hydrogen atoms are always in pairs with two H's side by side. Elemental oxygen atoms are always in clusters of various numbers of O's—commonly eight pairs of O's cling together in a fist of 16 oxygen atoms. When a hydrogen pair comes so close to an oxygen cluster that their atomic plus and minus electricity comes into play, an O darts out by itself and the pair of H's separate. Each hydrogen atom has one electron in orbit around its nucleus, which they now insert among the *six* electrons in the outer shell of the oxygen atom. This unites two H's with one O in an "electron sharing" bond.

That kind of bond makes a rather weak connection; such a molecule would be easily torn apart by colliding with another. However, the instant their electrons are in orbit together, the two H's and the one O find themselves firmly joined by the magnetic attraction of the negative charge of eight electrons orbiting the new H_2O molecule and eight positive charges in

---

*For interesting discussion of this subject see A. D. Moore, *Electrostatics* (Garden City, N.Y.: Doubleday & Company, Inc., 1968).

the nucleus of the oxygen atom. This, by the way, is the standard chemical bonding that makes all common compounds.

If that were all there is to an H_2O, it would not create water and this might be another elementary compound such as natural gas, ammonia, or table salt. But it happens that H_2O has two extraordinary features which make this little molecule unique on the face of the earth.

One is a terrific drive of H_2O's to attach themselves to each other by a special kind of coupling that is both stout and flexible. The oceans, lakes, rivers, and raindrops of our planet are the culmination of this olamic union of water molecules.

The extra strong bonds which unite the three atoms in a water molecule and cause H_2O's to couple eagerly to each other are comparable to the attraction of a magnet to iron filings. The strength of the tug is exactly in proportion to the distance between the positive charge of the magnet and the negative charge of the iron particles. When they come close they snap together vigorously. So it is with atomic magnetism, between orbiting electrons with negative charges and atomic nuclei with positive charges.*

Consider those hydrogen atoms attached to the big oxygen. Hydrogen is the tiniest atom of the universe. It has a smaller diameter than the atoms of all other elements. In short, each of those H's can get its nucleus (positive charge) *closer to* the electrons (negative charges) of an atom of oxygen than any other element. The law of magnetism gives this combination unassailable strength.

H_2O's, wielding this allurement, always unite with each other via their little hydrogen atoms. The O does not bind any water molecules. The positive pull in the nucleus of the oxygen atom is weakened where O's diameter is eight times bigger than that of H. This means that water is entirely fabricated by the extra strong *and flexible* couplings made by an H between two O's.

The water fabric holds an immense amount of dissolved matter, the atoms and ions of all the other elements. But these do not break into H_2O's or hurt their integrity. The water fabric

---

* On the molecular scale distance is measured in *millimicrons,* which act the same as fractions of inches in the case of the common magnet.

en masse in oceans and lakes holds a great tonnage of many molecules and particles of debris swept together by river systems from run-off on land. The heavier particles descend by zigzagging downward through the H_2O fabric *without tearing it,* ultimately forming sediments. However, *buoyant* things dance in the labyrinths of the H_2O's while, by dint of their hydrogen bonds, the water molecules remain intact and in good working order in the midst of the electrostatic excitement.

H_2O'S DEEP SECRET

The other extraordinary feature of a water molecule is the angle assumed by the two little H's when they affix themselves to the circumference of the big O.

On every hand we see beautiful geometry in nature, with water at the heart of it all. A drop of water possesses the urge to form a sphere. The dynamic spirals of the vortex of water are reflected in pine cones, fern croziers, the center of a daisy, snail shells, and waves tumbling on a beach. The circle of the ocean horizon is the most perfect geometric figure on the face of the globe. And everybody knows how neatly a circle divides.

So, one would suppose that an H_2O would have a symmetrical geometric pattern—for example, the two H's attached to the circumference of the O at a 45° angle. Or they might make a "four square" structure with the H's located 90° apart, a right angle. Somehow, 90° makes sense for a north-south pole design. Or perhaps the angle doesn't matter, merely the interplay of atomic energies in two exciting elements creates a water molecule—hydrogen atoms in capricious atomic tumult would simply collide and adhere haphazardly to an oxygen atom.

What actually happens is that electron alignments steer the H's to a pinpoint landing on the circumference of the O at a precise distance from each other—*always at an angle of 104.5°!*

That singular angle should be celebrated as *The Angle of Life.* It represents the crucial point at which the two most dynamic elements of the universe create a water molecule. This, in contrast to the ferocious gases and solid finalities of all other molecules created by raw elements—by virtue of 104.5°—made things happen on earth that could not happen without it. This "defect" in the harmonious geometry of H_2O puts it off balance electrically. It also makes the indestructible molecule restless and never satisfied.

The plus and minus signs in the electricity of two hydrogen atoms and one oxygen atom combined at this precise angle can never achieve placid pairs. They will combine with each other —negative electrons attracting positive protons in nuclei—but always there are unsatisfied charges left over to reach out and beckon to every molecule it meets. It is thought that the visible water of our everyday world is made of super molecules in which 7 or 9 H_2O's are combined. Yet those big *working molecules* continue to reach out with unsatisfied magnetism.

The permanent electronic restlessness that possesses water molecules, together with those tough, flexible hydrogen bonds, causes water to flow, erode, dissolve and mix other elements, run, rain, evaporate, and reappear as cloud droplets—while the basic H_2O unit remains intact. With that eccentric 104.5° angle of its makeup, water perpetually promoted the evolution of life on earth.

This is a quite recent discovery. The best reference I have seen which stresses its scientific importance is a book published in 1961.*

The precious angle eluded the pursuit of laboratory research because it is subtle and seemingly insignificant. In short, water is obvious, and bonding two atoms of hydrogen with one atom of oxygen seems to be just basic chemistry. Electrostatics is one of those scientific words for a natural phenomenon which seems to explain something, while in fact it only announces another mystery.

---

* Kenneth S. Davis and John A. Day, *Water, the Mirror of Science* (Garden City, N.Y.: Doubleday & Company, Inc., 1961). It tells complicated technicalities with clarity. A fine book for the layman who would like to get involved.

With their elegant techniques molecular scientists can only say *how* it works according to the laws of chemistry. But no man can say why those laws obtain—or, in connection with our subject, no one can say *why* that precise pattern of electrostatics creates an angle of 104.5° that transforms two gases into a fluid that made the surface of planet earth spring into life. Is it just a coincidence?

4

The Fabulous Fabric of H$_2$O

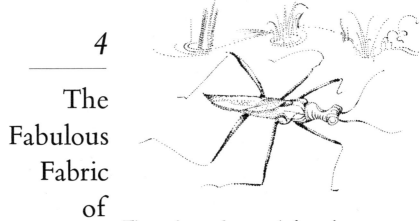

Three thousand years before the Christian Era builders discovered the remarkable structural possibilities of the pyramid form, and priests determined that pyramids have occult powers. According to Sir James Fraser in *The Golden Bough*, magic was rife in ancient Egypt and it is deemed that magician-priests supplied the fanatic zeal that built the great pyramids because they perceived that peaks pointing so emphatically straight up into the sky can exert influence on the gods. While today we scout the thought of magic, it cannot be denied that the Cheops Pyramid standing silent and motionless on the flat desert floor at Giza has cast a spell over people for 4,800 years by the majesty of utter simplicity and a sense of strength and timelessness.

The Greek mathematician Pythagoras visited the pyramids in 540 B.C. Pondering the sloping sides converging so exquisitely, he perceived that triangles and pyramids have remarkable properties. By interlacing three triangles a five-pointed star is made; by interlacing two equilateral triangles a six-pointed star appears—the pattern of snowflakes. Pythagoreans discovered arcane powers in triangles and their three-dimensional pyramids. To them the stars made by triangles betoken health and vitality and they avowed that a pyramid, the simplest figure that encloses space, is the first principle of the universe on which depend all its order and harmony—and the rhythms of music and poetry.*

Today, 2,000 years later, the views of the Pythagoreans about the mysterious powers of the pyramid figure do not seem wholly superstitious. Our molecular science is showing us H_2O molecules that have the simplicity of form of all pyramids, with unique strength and timelessness. Moreover, they give the water fabric remarkable properties, and they possess occult powers called electrostatics. The tides and vortices of the ocean reflect the harmony of the universe imparted by the spin and orbit of the globe, and waves have the rhythm of music and poetry.

But all comparison with Pythagorean pyramids ends there. Egyptian pyramids have square bases. This is the architecture of massiveness, stability, and permanence. Their only rhythm is that of proportion and line, and the only resonating with the harmony of the universe was in the imagination of the philosophers.

The pyramids formed by H_2O's have triangular bases—more accurately they have no base at all, for they are never in repose, but eternally tingling and whirling with electrostatic activity. It is their *energy* that is timeless, and by which they keep forever young and fertile.

In solid geometry the H_2O pyramid is a *tetrahedron—tetra* is the Greek word for "four." Each water pyramid has *four* faces of *four* triangles. A wire model has *three* wires meeting at each peak, and takes *six* wires to construct. What is more, a pyramid has a *center of equilibrium*, like a sphere. Indeed, pyramids and spheres are the same species. Any pyramid can form a sphere merely by whirling around its center of equilibrium. A pile of cannon balls or bowling balls makes a pyramid without any arrangement on your part.†

We catch glimpses of water drops forever striving to be spheres, and snowflakes always with six points or six sides.

---

* Samuel Colman, *Nature's Harmonic Unity* (New York: G. P. Putnam's Sons, 1912).

† The excitement of tetrahedrons, their relation to soap bubbles, and the way they generate spirals and symmetry in plants and animals is beautifully told in D'Arcy Wentworth Thompson, *Growth and Form* (New York: The Macmillan Company, 1942).

34

The restless water fabric is utterly dynamic due to that odd angle of 104.5° in the triangle made by the two hydrogens and the one oxygen. This makes an *isosceles* triangle, in which one angle is different from the other, hence two sides longer than the third. When these triangles combine in a water molecule they form a tall slender pyramid in which the angles and sides do not all match.

The lines of such a pyramid, as visualized in a water molecule, are *lines of force* between the *nuclei of the atoms*. Therefore, the corners of the pyramid represent the pinpoint locations of the positive charges of the atoms which exert the magnetic attraction that binds the H's and O's. The point is that since those lines are not all equal lengths they exert different strengths of energies. This disparity added to that of the great difference in the atomic energies of water's two elements—the positive charge in the nucleus of one oxygen atom is four times stronger than the positive electricity in the nuclei of two hydrogen atoms —commits H_2O's to be forever restless. This is the energy of water that generates running, swirling, and splashing.

We behold an incredible substance which seems so weightless in clouds, so transient in dew drops, so placid in the mirror of a pond—yet with power to convert a raw, lethal planet as a stage for life. Raindrops peppered mountains, eroding and rounding them to low hills. Running water in the Colorado River cleft the massive dome of the continental shield to carve the Grand Canyon a mile deep. Water seeping from snow on the Tetons in Wyoming flowed calmly across Idaho in the Snake River to carve Hells Canyon 5,500 feet deep through the basalt Columbia Plateau, hardest rock of the continent.* Delicate snowflakes in the Ice Ages, by sheer weight of volume from century-long blizzards, solidified as glaciers that planed off mountains, bulldozed the debris, and ground up the stones— then spread the sediments to become a soft blanket of soil on

---

* The energies of water are timeless because they are derived from the atomic energies of hydrogen and oxygen. Hells Canyon doubtless took millions of years to create. It bequeathed to us a peerless spectacle of the untouched splendor of our America. As this is written it is scheduled for extermination by the U.S. Army Corps of Engineers, and electric, power and light companies.

the continental rocks, and as a soft cradle of life on the floor of seas. We shall return to the Ice Age drama of water power preparing the lifeless globe to be the stage for life in Chapter 17.

Let us now visualize how the water fabric is woven by its pyramids and why it has an *open texture* which gives it such extraordinary properties.

A single H_2O reaches out with two plus signs from its hydrogen and two minus signs from its oxygen. The nearest H_2O's jerk to attention, bringing their opposite signs in alignment, which unites them by electrostatic attraction. It takes four H_2O triangles to make the basic tetrahedron of a water molecule in three dimensions.

No matter how many tetrahedrons stick together by paired plus and minus signs, a cluster can never reach stability. The peaks of peripheral pyramids continue to bristle and reach out with electrostatic magnetism. The molecular clusters vibrate, whirl, and dart—making and breaking contacts with electronic speed. This perpetual melee of H_2O clusters en masse gives fluidity to visible water. Yet water is ponderous enough to be the victim of wind and gravity, and can easily be torn to pieces, as seen in breaking waves, whitecaps, and all sorts of spray and splashing.

When water stuff is rent, it is repaired with electronic speed. You see raindrops on the windshield of your car literally snap together the instant they touch each other, and quickly form rivulets on the glass that struggle against wind and gravity to hold together. This, as we have just noted, is because H_2O clusters always have unsatisfied plus signs from their hydrogen and minus signs from their oxygen reaching out to find partners. Let us see how this electronics operates the great water cycle.

Warmth of sunlight on the blue ocean quickens the vibrating and darting of water molecules near the surface so much that some of them leap out of the sea and into the air, which they fill with water vapor. Although too small to be visible, their tetrahedrons are intact, their magnetism for each other is undiminished, but they have a violent urge to fly far apart. (This gives the expanding power of steam.) But when water vapor cools, the molecules slow down, draw closer, snap together, make droplets in visible clouds.

The amalgamation proceeds. Cloud droplets cohere in raindrops that fall to the ground and merge in trickles, that merge in streams, that merge in lakes, whose outlets deliver the molecules into the ocean where they came from.

The electronic attraction between H$_2$O pyramids preserved the water heritage of our planet through eras, giving water enough time to dissolve and assemble the elements for that unique unit of matter, a living cell. Thereafter, water, flashing its timeless electronics, caused that peculiar item to survive and multiply, kept it supplied with food energy, and disciplined its evolution.

This miracle performed by water—I challenge you to find a better word for it—entails subtle properties other than the kinetic forces of the global-water cycle. We detect these in the open fabric made by H$_2$O's. The "tall slender pyramids" cling together at their peaks, which represent the locations of the nuclei of the atoms, and as such they are the centers of attraction. Visualize pyramids attached at their peaks—they make a sort of *lattice* with space between each pyramid.

This is a gossamer fabric, indeed, vibrant, quivering, delicately elastic, highly sensitive, expanding and contracting instantly according to the minutest changes of temperature. Ceaselessly breathing gases and solutions, it can draw in, hold, and carry in its interstices all sorts of ions, atoms, and molecules of other elements without impairing in the slightest the wonderful water fabric.

Consider some of the properties of this gossamer fabric.

DISSOLVING POWER

In the beginning, when the first H$_2$O's condensed in steam meandering from volcanoes, and moistened stones in shady depressions, all the elements were either locked up in rocks or gyrating in turbulent gases. The earth was as cruel and lethal as the moon today, with one drastic difference—the earth had moistened stones and little puddles as soon as there were spots cool enough for H_2O's in liquid form.

A little water cycle started to operate with evaporation of moisture on the rocks and of the puddles, and then, in the coolness of night, more rocks became wet, and in the course of millenia the puddles became lakes, and the lakes joined forces in ocean basins. Most important was the force of dissolving.

From the beginning water became *solutions* of elements which it set free from the rocks—and which, without H_2O, would have been dead through cosmic time. Indeed, today chemically pure water is unknown—I do not know whether it can be made even in a laboratory if it contacts anything such as air or glass, so quick is its dissolving power. The water of life must be "polluted" by a diversity of vital elements which it has dissolved and carries in its fabric—carbon, nitrogen, silicon, phosphorus, sulfur, chlorine—and the minerals sodium, magnesium, potassium, iron, copper extracted from the rock vaults.

By the same token, those H_2O clusters of water vapor in the air have a big job of dissolving to do. Above all, those droplets must grasp with their dissolving fingers the carbon dioxide gas (CO_2) in the air. This other molecule made by two lively elements dissolves easily at the touch of water. The mixture of water and carbon dioxide makes carbonic acid, which descends to the ground in raindrops—all streams contain carbonic acid, which adds its electronic power to that of water in vitalizing the oceans. Were it not for this partnership with carbon dioxide, we can idly theorize it would have taken a billion years longer to make earth come alive—probably with only animated molecules!

The "dissolving fingers" of water are, of course, those magnetic plus and minus signs which make the tetrahedrons snap together. Water fabric of whatever size always bristles with this attraction, so it reaches out and joins nearly everything it touches. Electrostatics make it sticky—we say that water makes things *wet*.

The dissolving power of water is caused by a kind of atomic energy called *dipole moment*. That is a scientific idiom which simply says that H$_2$O molecules have a plus pole and a minus pole, like the North and South poles of earth's magnetic field. These poles of H$_2$O's exert a disrupting influence on all other molecules. They break up crystals—how fast crystals of salt and sugar dissolve before our eyes!—and thus release the elements, making them available for the molecules of life. Remember that the ions and atoms and other molecules like sodium chloride (table salt) are carried in the gossamer fabric in the lattice spaces between pyramids, without altering it—and they are given time in underwater collisions to make all sorts of fresh combinations.

Vital among these combinations is carbon dioxide in the air —as well as lots more of it dissolved out of rocks on the floor of the ocean. CO$_2$ and H$_2$O became inseparable companions in the atmosphere and in the ocean, always cooperating in the processes of life in two distinct ways. First, the carbonic acid they form together is a caustic chemical that gives more sting and piercing power to H$_2$O's magnetic dissolving, greatly speeding it up. The second way carbon dioxide helps water to carry on life came after organisms were created, and needed to eat.

As traveling companions in air and water, H$_2$O and CO$_2$ keep their identities. They do not combine to make any enduring substance—they do not compose any inorganic matter such as the rocks and gases of the primordial planet. But this energetic pair, with the help of sunlight energy, donate their atoms for the *carbohydrates* of life. They put their carbon, hydrogen, and oxygen into a special formula that is a source of the sugar and starch of foodstuffs, and structural materials for both plants and animals, such as wood, chitin for the shells of lobsters and shrimp, and the skin of mushrooms.

Moreover, as partners, H$_2$O and CO$_2$ work together in laying down limestone that installs a big reserve of carbon near the top of the crust where it is readily released by dissolving. Calcium is required for limestone, and great credit should go to the H$_2$O-CO$_2$ team for providing a bounteous supply of the mineral that makes the shells of snails and corals, clams and oysters—and the bones of vertebrates. Without dissolving—first to

39

lay down limestone, and then to release its carbon and calcium —life on earth would be puny and flaccid at best.

THE SKIN OF WATER

Usually we must leave it to scientists to tell us about molecu-large events, but the skin of water (formed by surface tension) exhibits visible performances that give everyone a chance to see H_2O pyramids in action. For example, boiling water is a dramatic exhibit.

Put some water on to boil, preferably in a transparent vessel. Watch and listen. Presently little bubbles appear, sticking to the bottom and low down on the sides. They are *air* bubbles, from air dissolved in the water. (Tap water has a good deal of dissolved air in it, especially after it has run and splashed out of the faucet.) The little bubbles indicate that molecules in water nearest the flame are so hot they are violently agitated, and their jerking and vibrating literally kicks out the dissolved air held in spaces between the pyramid clusters of the water lattice.

Now you hear a humming sound. This comes from little air bubbles so swollen with hot air that they let go and buoyantly burst through the surface. The hum is made by the breaking of the water skin; it has a high pitch because the air bubbles are little. But the water is not yet boiling.

Presently you hear crackling sounds. This means that water near the bottom has reached the boiling point (212°F) * so it must instantly evaporate, even though the water higher up is cooler. Now you see husky vapor bubbles (steam) rise, so it seems, out of the bottom of the vessel—only to vanish midway to the surface. This happens when the big, hot water vapor bubbles rise into a cooler zone, making them condense and revert to liquid water before reaching the surface. Under pressure from the weight of the water above they collapse violently, with reports muffled in the water so you only hear them crackle.

---

* No matter how high you turn up the flame, water never gets hotter than its boiling point. Water vapor carries off so much heat that it holds the temperature of the rest of the water steady.

These are reverse explosions (implosions) comparable to the bang of an inflated bag or rubber balloon suddenly collapsed.

Finally, the water is boiling from bottom to top. The surface is now a commotion of bursting domes made by heat-swollen steam bubbles pushing up against the surface tension, which stretches and resists mightily, but cannot curb the expansion power of hot steam. You hear the low rumble of what cooks call "a rolling boil."

Another fascinating exhibit of the skin of water is that of a slowly dripping faucet. I have a lazy habit of enjoying this when lolling in a bathtub. When there is enough water at the spout tip to fill its diameter it takes hold all around the rim. As water constantly backs up and presses on the water holding onto the rim, the elastic skin makes a bigger and bigger bulge, while a dome of water, seemingly so delicate and gentle, desperately tries to hold on. It dangles momentarily as though in a plastic bag—until too heavy for that gossamer H_2O to resist longer, gravity tears the drop away. But the skin of water is not torn—the falling drop is enclosed in a skin that snaps around it faster than the eye can see.

Like our polyethelene plastics, the skin of water is an arrangement of molecules, either linear or in circles—but unlike inventions of chemistry such as cellophane, the skin of water is exactly the same material as what it encloses. It is pure H_2O. Like the rest of water it is a commotion of energetic tetrahedrons. Indeed, some molecules are constantly quitting the skin while their places are taken by others. Moreover, of vital importance, many H_2O pyramids near the surface, where they are heated by sunlight, zip so fast they shoot right through the skin in the act of evaporating.

This leads me to speculate that the skin of water could have been a blueprint for the membrane of living cells—the vital regulator of the passing of particular solutions and ions of elements in and out of the cell. This started with the first drop of protoplasm which had to be enclosed by a membrane in the same way as a drop of water is enclosed by its skin. But there is a drastic difference—the skin of water allows *only H_2O's* to pass through.

That is why the ocean is salty, and raindrops are soft, fresh water. When warmth makes water molecules so energetic they rush through the surface skin of water and take to the air in evaporation, the sodium and chlorine of table salt are left behind to accumulate in the ocean through ages. This is also the case with the ions and molecules of many different elements. It is estimated that each cubic mile of sea water holds millions of tons of elements other than H and O. This includes gold, silver, and diamonds, and in some places on the floor of the ocean there are balls of manganese as big as baseballs.*

Today the desalination of sea water is a critical challenge, yet with all his vaunted technology man has never found a better way than nature's—evaporation using the skin of water as a filter. However, evaporating ocean water to meet the fresh-water needs of a population explosion is staggeringly expensive —both to apply enough heat energy and to transport the water any distance. Meanwhile, the superb water cycle is perpetually purified by sunlight sweeping the shimmering skin of water spread over two-thirds of the earth's surface, pulling up through it invisible clusters of H_2O's, that make fresh-water raindrops transported in flying clouds. It makes us look silly trying to tackle the challenge.†

If liquid water is just H_2O's, why is there a water skin?

This phenomenon is well described as "surface tension." I mentioned above that the pyramid clusters of water substance are turbulent, attracting and repelling each other with their opposing signs. Since these magnetic forces are exerted in all directions at the same time, the internal fabric is quite stable, with molecules equidistant and with pressures and pulls equal all over.

Not so at the surface, where molecules can be magnetically pushed and pulled by other H_2O's only from below, not from

---

* For description of substances left behind in the ocean by evaporation see J. L. Mero, *The Mineral Resources of the Sea* (New York: Elsevier Publishing Co., 1965).

† For a report about an alternative to desalination, with which saline water is used to raise crops, see Hugo Boyko, *Saline Irrigation for Agriculture and Forestry*, Proceedings of the World Academy of Art and Science, 1968.

42

above. This exerts a directional force that orients surface pyramids with each other, so they are *closer together*, increasing their magnetic attraction, giving the surface layer more tension and more strength.

The skin of water is one of the great wonders of the world. Only one molecule thick, it is so ethereal, so delicate that those evaporating, weightless H_2O's pass through it as though it weren't there. Yet the tensile strength of the skin of water is equal to that of steel. It would take the pull of a one-ton weight to rupture a column of pure water one inch in diameter!

This tenacious skin, so simply created by the hydrogen-oxygen hero, of our story, is always smooth, entire, flawless at every water surface—not only the surfaces touching air that we have been visualizing, but also everything with which water comes in contact. That directional force that makes surface tension would occur on a wet rock, on the underside of a puddle—and your body is instantly enwrapped with the magical skin of water when you dive into a swimming pool.

An engaging thought is that a replica of every submerged object is wrought by water skin. In the case of a ship this would be a model of the submerged part of the hull, precise in every detail. The weight of water displaced in creating this model is the weight of the whole ship—properly called its "displacement." This profound fact, that the volume of water pushed aside by a body submerged in it is the same weight as the body, was discovered by Archimedes while taking a bath at Syracuse, Sicily, around 200 B.C. When the idea struck him, he leaped from his luxurious marble tub crying, *"Eureka!"*—Greek word for "I have it!"

Some fanciful tribes of little creatures with Greek names inhabit the two-dimensional water skin world. The *Gyrinidae* (gyraters) scare off predators, such as hungry birds, by churning the surface madly like outboard motor boats that have lost their rudders. They keep an eye out for things to eat both above and below the surface at the same time with dual eyeballs —the upper hemisphere sees in the air, the lower under water. When these whirligig beetles dive they carry along an air bubble to breathe, like spacemen.

43

The *Notonecta* (back-swimmers) lie on their backs on the water skin with six legs spread out horizontally like oars at rest. When prey caught in surface tension floats within reach, those oars flash and grab it. Then the water boatman rows off fast to a hiding place to enjoy the catch for dinner.

Halobates (sea-haunter) is a wingless relative of the water striders which spread wide their six long legs, reducing to one-sixth the pressure at each point, and thus they do not pierce the water skin—they skate around making dimples in it. A care-free tribe of water striders has recently been discovered in mid-Pacific, a thousand miles from land, where they stride on the smooth skin of ocean rollers, breeding, laying eggs on bits of seaweed, and drinking solutions of dissolved organisms that spread soup on the water skin. It is an area of doldrum calm where the ocean surface revolves slowly in a huge whirlpool, accumulating the surface fertility. An occasional breeze may make a splash that knocks the sea-haunter under water—its body is covered with a fine velvet pile so under water it is clothed in silver bubbles, with which it bobs up serenely.* These inhabitants of the water skin world live well-fed and secure, thanks to a remarkable property of H_2O.

However, the greatest work of the water skin in life is capillarity. *Capilus* is the Latin word for "a hair" and refers here to the tensile strength of a thread of water in a very fine tube. In this situation a high percentage of molecules are in contact with the sides of the tube, which orients them in surface tension, and this imparts strong cohesion to the whole column of water.

Since H_2O pyramids are ever energetic, a remarkable force is exerted along the length of the thread of water enclosed in the tube. Those at the upper end—if we visualize a vertical tube—grasp the sides strongly with wet, sticky, electric fingers, while the electrostatic attraction of the tube itself pulls on them and they keep grasping higher and higher as they climb the wall—pulling the column of water up and up. If the thread of water is

---

* The mid-Pacific water haunters are not mentioned in any book I have consulted, but for other diverting accounts of insects see Howard E. Evans, *Life on a Little Known Planet* (New York: E. P. Dutton & Co., Inc., 1968).

44

continuous it does not stop rising in the tube until its volume of water, elongating as it creeps upward, becomes so heavy that gravity calls a halt.

This dynamics of water moving (not only vertically but in every direction as well) is at work on every hand in our living world. The finer the tube, the greater the percentage of surface tension, the stronger the pull, the higher it goes. It raises groundwater 4 feet or more through the soil, where the capillaries of root hairs accept and pull the water into a tree trunk. It mounts a few feet higher in the porous wood, and from there on, the strong pull of vacuums in the canopy where leaves are expelling water into the air pulls the water to the top of the tree.

Capillarity is highly convenient for us. The minute blood vessels called capillaries deliver blood to every distant cell of our bodies. We are clothed in a fine-spun capillary garment with the same sort of molecular forces as those in the skin of water.

A NEW FRONTIER OF KNOWLEDGE

We are prone to look back at the milestones of human awakening—Archimedes in his bathtub, Newton's laws of gravity, Franklin with his kite, Darwin's finches at Galapagos, Pasteur's microbes in the air—as though all the big news about our predicament on this planet is in the past. However, as this is written, the "atmospheric sciences and environment" are in full swing blazing trails of fresh discovery.

That this is impelled in part by horror at the pollution of water and air (see Chapter 18) makes reports from this area no less epochal. What interests us here is the interplay between the energies of water and energies of organisms. This is dependent entirely on the dynamics of close-packed H₂O's in the skin of water.

The "standard" water fabric within its skin, in ocean, river, or raindrop, acted for two billion years (and still does) as elemental, inorganic H₂O pyramids operating the water cycle, and maintaining temperate climates by carrying warmth around the globe in clouds and ocean currents. After the miracles of protein and chlorophyll, every organism and the evolution of life employed the subtle forces of surface tension.

At the start of the 1970's exciting news about this peculiar phenomenon, water skin, is coming from both the atmospheric sciences and oceanography. This concerns *buoyancy* of boats and fish; the friction of water flowing past the hull of a ship trying to go as fast as possible; seeding clouds to produce rain; ways to dispel fog; news about the reactions of air, water, and sunlight in photosynthesis; and dramatic new discoveries about lightning and thunder. How can small gentle raindrops stage a flash of lightning that exerts hundreds of millions of volts? Science is taking a fresh look at the skins of raindrops. We must leave these big subjects to others *—with only a brief mention here of two items from the new frontier of knowledge that underscore the eerie energies of the skin of water.

Polywater was hailed in 1969 with bedazzled reporting. Headlined as "unnatural" water, it was regarded with awe and fear like those imaginary microbes on the moon. It had been discovered and described by a Russian scientist a few years ago, but western scientists didn't believe the reports. A jittery physicist, questioned about polywater, was reported in a leading national weekly as saying: "Once it is let loose, the stuff might propagate itself, feeding on natural water. The proliferation of such a dense, inert liquid could stop all life processes, turning

---

* A few good references:
Duncan C. Blanchard, *From Raindrops to Volcanoes* (New York: Doubleday & Company, Inc., 1967).
W. H. Westphal, *Physics for You and Me* (London: George G. Harrup & Company, Limited, 1962).
David M. Gates, *Energy Exchange in the Biosphere* (New York: Harper & Row, Publishers, 1962).
A. D. Moore, *Electrostatics* (New York: Doubleday & Company, Inc., 1968).
Reginald C. Sutcliffe, *Weather & Climate* (London: Cox & Wyman, 1966).
Frank W. Lane, *The Elements Rage* (Newton Abbot, England: David & Charles, 1966).
Bernard Vonnegut, State University of New York at Albany, "Thunderstorm Electricity," *Discovery*, Vol. 26, No. 3 (1965), 12–17.
Publications of The International Oceanographic Foundations (Virginia Key, Miami).

the earth into a reasonable facsimile of Venus." A researcher at the National Bureau of Standards, after studying polywater with the most sophisticated techniques, said that "until science knows more about polywater, it should be handled with care."

What makes polywater spooky is that it is just water—its formula is H_2O—that doesn't obey the rules. Here is a container with 100 pounds of water; beside it is the same size container with the same volume of polywater—the latter weighs 140 pounds. The haunted water doesn't boil until it is 780°F; it refuses to freeze until it is 104°F *below zero*. People shaken by what happened when science tinkered with forces hidden in the nuclei of atoms see disaster in polywater.

Let us speculate about this in terms of those H_2O pyramids that are pulled closer and oriented by their magnetism when the attractions of other H_2O's pull in only one direction—down, in the case of surface between air and water—instead of all directions at the same time. We have just seen how surface tension is a bigger proportion of water in a slender tube—exerting the vital force of capillarity.

Polywater is produced by H_2O molecules confined in a microscopically fine glass tube. Those pyramids grasp the glass as in capillary attraction—but this tube is only a few husky H_2O clusters in diameter. There is little or no internal fluid. This is a tight filament, with all the energies of the pyramids constrained to pull in a linear direction, with no relaxation, no chance to dance.

This makes a peculiar electron energy pattern—perhaps all the more so if the glass that squeezes so tightly joins its electrostatic influence to that of the H_2O's—which makes the glass influence the water molecules like a catalyst. The H_2O pyramids become distorted in this unnatural situation. Hydrogen atoms form two bonds with one oxygen, but internal energy is the same as in normal water—and polywater can never form in nature. It is merely a temporary tight squeeze in a laboratory.*

Incidentally, nature seems to have been making use of "poly-

---

* This is based on a summary of a report by Princeton University chemists, which dispels the notion of a "new kind of water," published in *New Scientist* (London, England, March 26, 1970).

power" inside the cells of those very scientists who regard poly-water with trepidation. A scientific journal reports the discovery of a counterpart of polywater in protoplasm, where it increases vital viscosity and bestows other peculiar powers. It is identified as that weird, "new kind of water" like polywater.

The tightly linked chains of H_2O pyramids not only make protoplasm viscous, they also make it slippery. For slipperyness nature gave fish—and particularly swift swimmers such as squid, and smooth, graceful swimmers such as dolphins—skin glands which exude a viscous lubricating fluid that reduces surface tension friction.

Industrial research has found ways to give ordinary water slipperyness by orienting H_2O's with chemicals, without going to extremes with fine glass tubes. Slippery water was recently ceremoniously bestowed on the New York City Fire Department. With less friction, slippery water travels faster and the same volume of water can be delivered with smaller diameter, lighter hoses. Firemen are able to climb stairs and reach remote locations more rapidly with less weight to carry.*

WHIRLING DOWN THE DRAIN

Why the beautiful spirals of water going down the drain of your bathtub and kitchen sink? For the same reason that streams meander. Elastic, flexible surface tension makes water stuff describe graceful curves. Pyramid molecules cohering at their peaks make a fabric that can sway and bend but not turn sharp angles. When its flow is deflected from a course it must sweep around. Once started on a curving course it holds that direction until deflected by some influence—for example, pipes straighten out spirals that disappear down the drain.

It is said that water going down the drain revolves in a specific direction in the northern hemisphere and counterwise in the southern hemisphere. I have even seen the question asked in a scientific journal whether a passenger on a ship crossing the

---

* E. H. Blum, "Slippery Water in Fire Fighting," The Rand Corporation (New York, June, 1969).

equator might see the vortex of his bath water pause and change direction at latitude zero.

The illusion rests on a huge geophysical fact that the spin of the planet deflects the direction of flowing water. Looking from the North Pole the earth's surface drag is from right to left, counterclockwise. Looking from the South Pole the drag is from left to right, clockwise. This would cause water flowing freely on the surface to be deflected—starting it on a counterclockwise spiral course in the northern hemisphere, and clockwise below the equator. This is called the *Coriolis* force. It is not inertia, such as makes a sailboat sheer and haw. It only affects a thing in motion, trying to go in a straight line across a revolving surface.

Try this. Put a cardboard in place of a record on a record-player turntable. Start it revolving and then without paying attention to the turning motion, draw a straight pencil line across it. You find yourself making a dynamic curve worthy of an artist. If you draw the line from the center to the edge in any direction it creates the spiral of a chambered nautilus—and of the vortex of water going down the drain in the southern hemisphere.

There is no doubt that the spin of the earth influences ocean currents. For instance, an equatorial current of the Atlantic was deflected northward by the Coriolis force to become the Gulf Stream; and it makes the surface of the Florida Current slope up so much that water level at Cat Cay, Bahamas, is two feet higher than at Miami. But does revolving on the grand scale of sunrise and sunset affect a miniscule of water going down a bathtub drain?

To solve this puzzle, the Massachusetts Institute of Technology set up an experiment. A circular tank, 6 feet in diameter, 6 inches deep, with a drain hole in the center ⅜ inch in diameter, with flat horizontal bottom, was utterly isolated from air currents, sound, vibration, changes in air pressure, and temperature. Nothing but the spinning earth could affect the motion of the water in the tank as it drained. It was filled with a jet that gave the water a circular motion, clockwise, that is, contrary to the Coriolis pull in the northern hemisphere.

It took 24 hours for the water to lose all momentum and become perfectly still. Then a little cross of wood slivers, weighted by a tiny wire cut to a precise length to be buoyant just beneath the surface, was placed over the drain hole. They watched through instruments. In fifteen minutes the indicator started to rotate counter-clockwise. When the tank was almost drained it was making a complete revolution in about three seconds.*

So—it is a scientific fact that the spinning of the planet tends to cause water going down a drain to revolve in one direction in the northern hemisphere and contrariwise in the southern hemisphere. But, in our everyday world this cannot be detected! In real life, water drains with a whirl in either direction, by chance. It all depends on the original impulse from a faucet, or the slope of the porcelain at the drain hole of the tube, or a slight deflection encountered.

As for the great ocean currents, they are greatly influenced by promontories, the contours of a shore, depths of the sea which determine friction on the bottom, and the wind and sun, as well as by the planet's spin.†

---

* A. H. Shapiro, *Bath-Tub Vortex* (Cambridge: Massachusetts Institute of Technology Press, 1962).

† The drama of the Gulf Stream and other great rivers of the oceans and their effects on climate, the tides, and storms are beyond the scope of our book. There are many excellent books, among which I particularly recommend: Henry Chapin and F. G. Walton Smith, *The Ocean River* (New York: Charles Scribner's Sons, 1962); Pier Groen, *The Waters of the Sea* (Princeton, N.J.: D. Van Nostrand Co., Inc., 1967); Richard Carrington, *A Biography of the Sea* (New York: Basic Books, Inc., 1960); P. H. Kuenen, *Realms of Water* (New York: John Wiley & Sons, Inc., 1963).

When fresh water aboard ship is low and stale, sparkling puddles on the ice pans look as tantalizing as the purest spring water.

The open ocean is a composition of hydrogen, the element that is 90 percent of all the matter of the universe, and of sunlight that is the energy of the solar system. Pondering it you have intercourse with infinity.

Explosions of ocean waves on a rocky coast make the continental rocks tremble. An impact power of 6,000 pounds to the square foot has been measured for a hurricane wave.

The argonaut shell is an exquisite egg case, so light it is a bobbing little sailboat for the argonaut babies.

Looking down through the surface of low-tide pool we catch a glimpse of life in the seaweed jungles.

The incoming tide splashes into a pool delivering bubbling nourishment from the open sea on an exact timetable.

The towering bow wave engulfed the ship two seconds after the picture was taken. Terrible is the power and tenacity of H_2O's driven by wind.

In Bay of Fundy tide may rise 50 feet, comes in as a big wave called a "bore," suddenly covering this mud flat teeming with burrowing creatures.

People who built their houses by estuaries have enjoyed riches of clams and crabs, fish and lobsters.

5

Water
Assembles
Life

In 1965 I witnessed an event in the laboratory of the Institute for Space Sciences, University of Miami, Coral Gables, Florida, which had all the earmarks of a magician's trick. It was a scientific demonstration of how life *could have* been born on the primordial planet among volcanic debris, when the atmosphere of the raw globe consisted of ammonia gas and methane (natural gas), and had no free oxygen. The origin of life under these conditions has posed a baffling problem for biologists.

In pursuit of an answer, 34 of the highest authorities of our day met in Florida to argue the question under the auspices of the National Aeronautics and Space Administration. The panel included scientists of the caliber of Oparin from Russia, Haldane from India, Bernal from the University of London, Mirsky and Lipmann from Rockefeller University, New York. The record of the proceedings was published in a book, which is an extraordinary science detective story.*

In his contribution, Dr. Lipmann likens planet earth to a living organism with elemental energy cycles that are in themselves unrelated to life. He points especially to the water cycle that converts the heat of the sun into mechanical energy—released

---

*Sidney W. Fox, Ed., *The Origin of Prebiological Systems* (New York: Academic Press, 1965).

in raindrops, waterfalls, river currents, and ocean currents. He stresses that life developed wholly dependent upon this energy cycle. According to Dr. Lipmann the energy cycle inside living cells is a simile of the global water cycle.

If some link can be found between the wild, violent water cycle and the exquisite river of life inside a living cell, it could be interesting evidence of how life began. The link might be a chemical formed by elements dissolved by water from raw mineral ore such as lava. Is it possible, under laboratory control, to produce molecules that will cling together in chains like protein, or resemble bacteria—or act lifelike when they are not alive at all? The event I witnessed at the University of Miami answered *yes* to that question.

They set the stage to duplicate a spot on the cone of a volcano, high up near the lip of the crater where sizzling lava stones fall. One of the scientists flew to Hawaii and picked up scoria freshly spewed from a fire pit while the stones were still warm and bristling with the same minerals as those of the crust of the pre-life planet. This material, carefully protected from contamination, was brought to the Florida laboratory.

They distilled amino acid (the vital chemical of protein) out of methane, ammonia, and water vapor—the gases that constituted the atmosphere of the pre-life planet—then dunked a hunk of the lava in the amino acid and put the concoction in a laboratory oven to bake at 338°F. That was the mean temperature about 4 inches below the surface of the lava gravel that forms the cone of an active volcano, where rainwater tends to concentrate, soaking the gravel while not evaporating rapidly.

The material was left to bake for an hour or so while we ate lunch, and while the mineral elements of the lava and the carbon-oxygen-hydrogen-nitrogen of the amino acid made a hard, dry compound. This was allowed to cool slowly while being doused with water imitating heavy rain on the pre-life planet. The result was to give the lava a black sticky coating. A bit of this was put under a microscope at × 900 magnification, and one of the scientists, with a flourish and a smile, invited me to take a look.

I shall never forget what met my eyes. Instead of formless goo

I saw a galaxy of translucent spheres, sparkling in the spotlight of the microscope. They were jiggling, rushing around, bumping into each other—and particularly interesting was the way some of them clung together in beadlike necklaces.

The lively spheres were about the same size as *cocci* bacteria, the most primitive of the bacteria. Their performance of clinging together in slender chains is exactly like *blue-green algae*, the most ancient type of chlorophyll-bearing plants on earth today. The earliest trace of life on earth is the fossil of a blue-green alga in a granite pebble from Southern Rhodesia, shown by radio-chemical dating to be 3.3 billion years old.

It is tempting to speculate, as did the scientists who set up the experiment, that these chemical forces in volcanic cinders steaming in thunderstorms on the primordial globe initiated molecules of life.

Of course, the sparkling, necklace-forming spheres which I saw summoned by water out of raw lava and gases were not alive. They had no nuclei, no cell organelles at all. Joining end-to-end in chains is "polymerization" of everyday chemistry by which little molecules unite to make big molecules. However, is it just a coincidence that protein, the key molecule of life, gets its astonishing powers and agility by linking its amino acids in long "peptide chains" which bend, fold, and spiral in live cells? Is it just a coincidence that fourteen of the eighteen kinds of amino acids of protein compose those alert spheres from the soaked lava?

They are called "microspheres" in the report, which argues that they are "cell precursors"—that is to say, chemical droplets from which living cells evolved. They certainly show what happens when amino acids distilled from the gases of the planet's first atmosphere combine with elements from lava. Lively microspheres, the same size and shape as bacteria, appear! Note that elements in the microspheres dissolved from the lava include chlorine, sulfur, phosphorus, magnesium, which play dynamic roles in life.

When Dr. Oparin from Moscow, author of the classic Oparin Theory of the origin of life, saw microspheres, his comment was, "Since they are so beautiful one has to be especially critical in their evaluation." If the celebrated Russian biochemist was dismissing microspheres with a professional smile, why not? They cannot be evaluated even as a rough primitive model of living cells. But I think that they are exceedingly interesting as creatures of natural laws of physics and chemistry.

The famous Oparin Theory was that life began in "dilute soup" rich with organic molecules that thickened in warm sunlight to become jellylike (colloidal) protoplasm. This was assumed to have occurred in a nook of a low-tide pool or estuary where the broth of life could simmer undisrupted by eddies and splashing while receiving fresh supplies of vital elements from the ocean or dissolved from continental rocks.

Biologists have played fascinating games with theories about the origin of life, some calling it a sudden event that was incredibly unlikely. But whatever the variations of the Oparin Theory, there has never been any doubt that life was created in and by water. And there is towering evidence that, following the "dilute soup" stage, the items of life stuff needed a huge ocean, with vital dissolved minerals, in which to multiply through millennia and become a permanent feature.

The life of the rudimentary ancestors of blue-green algae would have been extremely tentative in a low-tide pool or estuary, and above all in a fresh-water pond. (As I write this, a scientific journal surmises that "alternate freezing and thawing of a fresh-water pond" was the mechanism that created life.) The cataclysms that assaulted the globe in eras after life appeared were appalling, according to the geologic record that emerges from the black depth of time. The rock bones of the planet betray the awful agony of its crust with incessant volcanic upheavals; orogenies lifted mountain ranges five miles high, then they were eroded flat and were lifted again; colossal cracks in the basaltic sima of ocean floors show how the crust split and land blocks drifted around the globe; and the ocean tides rolled over large proportions of the continents. After drops of chemical solutions achieved the status of living cells, it was essential that they reside in the ocean if they were to survive through millennia while they established lines of descent.

Oparin was right that microspheres are "beautiful," and everyone agrees that they are not alive in terms of being capable of self-maintenance, development, and reproduction. I regard microspheres as a beautiful example of Dr. Lipmann's elemental energy cycles on the pre-life globe. They show us an important happening when water went to work on the juvenile planet—dissolving atomic energies from lethal gases and from fresh hot lava, and reshuffling those energies.

Consider how organized and vigorous atomic energies became in a combine the size of a living cell. Consider how cocci bacteria and blue-green algae reflect their urge to rush around and cling in chains. This stretches the mind groping for the origin of life. It says that when sunlight, water, and certain restless atoms of the universe met on our planet they brought forth life.

THE PROTEIN MIRACLE

All phyla (lines of descent) of plants and animals in the world were created in the ocean. Things akin to algae and bacteria are transformed into diatoms and seaweeds. One-celled animals turn up with whipping hairs for swimming about and transform into the elegant *Paramecium*, whose single cell can navigate and hunt for food and reproduce sexually. There are countless kinds of shrimp, there are crabs and bizarre lobsters. At some point, in the dynamic salt water, fish appear. Eventually, evolutionary forces, set in motion under water, develop momenta so that life leaps out of the sea—and there are amphibians, reptiles, birds, and mammals.

Such awesome evolution called for perpetual reshuffling of elements and molecules, new combinations of atomic energies, ceaseless fresh supplies of hydrogen, oxygen, nitrogen, carbon, etc. That requirement could be met only in oceans that enwrap the globe. No traces of water vapor in an atmosphere, no wisps of moisture in dew or frost, no epochs of cloudbursts, not even lakes and rivers which are all temporary in geologic time, could have evolved any such marvels as fish, trees, insects, animals, or even "little green men"—at least, not according to the universal laws of physics and chemistry. Such things could only have evolved in oceans.

The word "miraculous," long taboo among scientists because of its supernatural echo, is heard today where scientists probing the depths of living cells are filled with wonder. In a way, protein, the giant molecule of all life substances, is more prodigious than, say, Aladdin's lamp. The latter had to be rubbed to produce one miracle at a time, but protein summons up countless miracles all at once with electronic speed. Some conspicuous ones are *chlorophyll*, the green molecule that traps sunlight and turns it into sugar; a *mitrochondria*, the phosphorus powerhouses of cells; *DNA*, the coiling tapes which record the inheritance code of every kind of life; *enzymes*, messengers that carry DNA orders in cells. Things like these are as "miraculous" as any legendary miracle.

Protein is not itself alive. It is just an extraordinary chemical. It does not exist in the raw realms of nature—only in a living cell. Moreover, no protein can be made in a laboratory starting from scratch, although the chemical makeup of protein is known. For instance, a protein molecule of cow's milk has 1,740 carbon atoms, 2,819 hydrogen atoms, 550 oxygen, 440 nitrogen, and 19 sulfur! When scientists mix those precise proportions of those elements in a solution and stir and stir and stir, no trace of protein turns up. Perhaps protein might be created if the laboratory assistant stirred the solution for a hundred million years?

The count of atoms for the cow's-milk protein molecule never varies by a single atom. Moreover, each protein molecule must have a mineral atom attached to it at fixed intervals. The selection of the mineral determines the function of the protein—whether it makes a muscle, egg white, the germ in a wheat grain, etc. For example, with an atom of iron, protein makes red blood for a mammal; with a magnesium atom, it creates chlorophyll for a leaf; an atom of copper makes the blue blood of a crab; cobalt makes protein for a lobster; nickel is chosen for mollusks; and lead serves the protein in a variety of marine animals.

The fantastic lengths to which life resorted in the creation of the phyla in the ancient seas is seen in the sea squirt, which manages to collect rare vanadium atoms to make its body fluid. Sea squirt also shows us how, in the primitive stages of evolution, organisms used trial and error. When no vanadium can be found, this animal substitutes chromium, or tungsten, or titanium delivered to our planet in meteorites. In a later chapter we shall have more to say about the ugly, interesting sea squirt, which shoots out water like a water pistol.

If ever there was an "impossible" molecule it is protein, an extremely long slender chain studded at regular intervals with jewels of copper, magnesium, iron, or other minerals. With such large but precise numbers of atoms in the basic protein molecule, and such a critical choice of accompanying minerals, how could such a molecule ever occur in the ancient ocean?

Today's computer mathematics provide a concept for the origin of protein. They explain atomic events and diagram the way molecules unite and break apart in millionths of a second, by using everyday numerals with little numbers (called a "power") that multiply them far away into infinity. With computers, according to the law of averages, the "impossible" becomes possible.

In today's space sciences this concept is called a "probability." This term does not mean that something is likely to happen within human experience, but that it *can* happen given an infinite number of chances. The cliché is that of a pack of cards flung into the air an infinite number of times eventually must fall to the floor in the order of the suits.

Peer into that profound blank before the blue-green alga that left its fossil in the Rhodesian pebble was created. Here is a situation in which the law of averages enjoys three infinities: *Time*—the ocean of the watery planet is timeless. *Space*—for atoms the ocean is as infinitely vast as sky space. *Numbers*—the collisions of electrified particles in a drop of water are in the order of "billions" per second.

In the ancient ocean the ions of the elements washed together in currents, splashed together in surf and swirled in whirlpools in mid-ocean. As they zipped and collided in the turbulence, they combined to form acids, alkalis, and salts with their magnetisms, which certainly created many "impossible" new kinds of molecules. These would have been evanescent, disappearing in millionths of a second, never to be a reality, except one that interests us—a rudimentary protein!

This is imaginary, but not metaphysical, and it relies on the "probability" of modern physics. This unlikely molecule, to survive, must have found itself in an environment like Oparin's dilute soup. Then with its unique powers of attraction and influence it gained in strength and size before restless water could sweep it away. At the same time, helped by other elements that crowded around, it was fenced in by a membrane, became an organism. Thereafter, the development of internal structures may have taken countless epochs. From that time to this day living cells enjoyed a teeming privacy in the ocean, and later in the sap of plants, and bloodstreams, which are arms of the sea.

THE FIRST POPULATION EXPLOSION ON EARTH

Protein went right to work to shape up life even before there was free oxygen either in the air or in the ocean. The first inhabitants of the planet eked out their lives by extracting energy from iron, copper, and sulfur which their ocean water dissolved from rocks. Certain algae bacteria called *anaerobes* (non-oxygen-users) still subsist on this austere staff of life, which needs only salt and carbon dioxide, plentiful in ocean water, to convert into energy. But note that the iron-copper-sulfur eaters appear not to have evolved at all after some two billion years. If water had not assembled a certain miraculous molecule, life on earth would have been nothing more than microscopic specks in the ocean.

This magnificent molecule—chlorophyll—changed the whole future of life by pouring free oxygen into air and water. This innovation, this triumph of protein, makes life a double miracle. It happened in the single-celled algae in the ocean deep, turned them green, caused them to initiate photosynthesis, by which life on earth could pack the energy of sunlight into food.

58

Chlorophyll is just as important for our living world—and just as "impossible"—as a molecule of protein.

A chlorophyll is an elegant structure built with 55 carbon atoms, 72 hydrogen, 5 oxygen, 4 nitrogen, and 1 magnesium. This chlorophyll marvel seems to have turned up quite soon after the iron-copper-sulfur eaters—perhaps it was a matter of but a few million years. The "probability" concept was again at work creating this second impossible molecule, where the anaerobes swirled in a milieu of infinite electronic events bonding and breaking infinite combinations. But now chemical reactions had a vital quality, with the chief elements of life concentrated in the "soup" that cradled the first living cells, and, moreover, a new organizing force was now operating—organic evolution, survival by natural selection.

The great shift was from reliance on static energies in rocks to radiant energy delivered by sunlight. The original source of energy from minerals was slower, harder to obtain, and localized. The swirling anaerobes had to be near their iron-copper-sulfur sources, and they had to wait until water dissolved these elements tightly locked up in hard crystals. What a dazzling revolution was wrought by chlorophyll! Sunlight saturates sea and land with the rhythm of the whirling globe.

The one-celled organism in the ancient tellurian mixing bowl —which assembled the right combination of hydrogen, oxygen, and nitrogen atoms and, by adding a single atom of magnesium, transformed it into a chlorophyll molecule—flaunted new power with endless possibilities. Doubtless this was the first population explosion on earth. The multitudes of tiny green cells, suddenly independent of the dissolving rocks, multiplied in upper layers of the ocean, where they spread over the world in windblown currents. Now life swarmed in sparkling waters, no longer lurking and listless in dark rocky lairs. As they danced in the sunlight, the photosynthesizing organisms launched the fabulous evolution that would lead to many-celled, elaborate seaweeds, jellyfishes, and sponges, and onward to glorious trees, birds, and animals in a future epoch.

This critical turning point of life on earth, which gave evolution new dimensions, was due to the shift in the source of energy. The iron-copper-sulfur eaters didn't store any food for future use, nor did they share their mineral rations with others. Their energetic ions (electrified particles) were instantly used up, and each cell was on its own. Chlorophyll packages sunlight energy in sugar or starch that can be stored within, between, and around the cells of a body and hold the energy for future use, ever ready to be released by a touch of oxygen. When a great surplus of food is stored up by chlorophyll, far beyond the personal needs of the cells which make it, organisms with no chlorophyll can get energy by devouring the green ones. The devourers are the members of the Animal Kingdom!

Water in photosynthesis plays its mighty role of transporting elements and supplying the right temperature for chemical reactions, and moreover, it puts the hydrogen of its H_2O molecules into the packaged sunlight. We have noted that the water molecule is indestructible under ordinary conditions. It cannot be rent by freezing, or evaporating, or when dissolving crystals. Yet quiet and motionless chlorophyll splits H_2O's! (Elsewhere this occurs only in the lethal power of ultraviolet radiation, and in flashes of lightning that strike with 18 million kilowatts of electricity in a thousandth of a second.)

The hydrogen atoms torn away from H_2O's by green chlorophyll are installed in sugar or other carbohydrates, and the atom of oxygen is thrown away to be free in the atmosphere. Thus, in the course of time, water, by assembling chlorophyll in the ancient seas, changed earth's original poisonous atmosphere into breathable air, long before the evolution of land life.

6

The Womb of Life in the Ocean

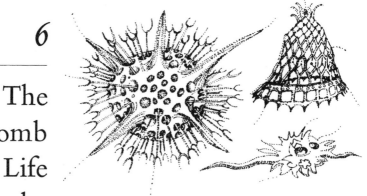

Out beyond where ocean rollers start their rush toward the shore, from just under the surface to a depth as far as sunlight penetrates, the most ancient and populous community on earth rises and sinks with the rhythm of day and night, swings as fluent as the ocean currents. This community is aptly called "plankton," from a Greek word meaning "that which is made to wander and drift." It has wandered and drifted in ocean currents, circling the globe through geologic time.

Because plankton plants and animals are pinhead size and hidden under the scintillating blue surface, they were not discovered until "the use of the townet and microscope in the study of this multitude of tiny drifting plants and animals of the sea" around 1850. Thereupon, oceanography was enlivened with excitement like that of discovering weird beings on another planet.*

Plankton is not a blanket of life evenly spread under the surface of the open ocean. It is denser in some areas, as around the British Islands and on the Grand Banks of Newfoundland where the fishing is great. However, drifting in ocean currents, stream-

---

* Early explorations of the ocean depths are a thrilling chapter of history. They are narrated in a superb classic: H. U. Sverdrup, Martin W. Johnson, Richard H. Fleming, *The Oceans* (Englewood Cliffs, N.J.: Prentice-Hall, Inc., 1942).

ers of plankton curl around the globe. The same kinds of plankton life are found in all the oceans. Yet above one-celled algae
and bacteria there is no foreordained shape or size of life, but
dazzling diversity.

About 500 million years ago, at the beginning of the Cambrian
period, some elaborate organisms became entombed in sediments
that hardened to rocks, and we find them today as impressions of
worms, centipedes, various shrimplike creatures, and primitive
shellfish—the earliest many-celled organisms in the fossil record.

Our interest now is what plankton tells us about events in the
black well of time which geologists call the pre-Cambrian, when
the lines of descent, the phyla, of life were organized. The most
arcane and timeless era of the evolution of life on earth can now
be studied in all its color and wonder with the instruments of
today's space science that observe atoms, electrons, and radiation.
This is because the pre-Cambrian drama of water and life is
still being played in the same way, with the same natural forces,
in today's ocean, where geologic time somehow loses importance.

Explorers of the plankton world can witness the actual events
that created those elaborate organisms that staggered across the
beaches and started our fossil record. The plankton community
is still enacting the exuberant Pre-Cambrian Prelude, in the original setting of radiant sunlight and elemental forces in the
swirling sea.

Ocean water is primarily a mixture of strong solutions of
hydrogen, oxygen, nitrogen, and carbon. In addition, sea water
is charged with energetic elements essential in all the organs of
cells and structures of life. Among these, *iron* is of prime importance. Without this magnetic element there could be no life
on earth as we know it. Iron kept the original "iron-eating"
bacteria going, and solutions of iron in sea water were on hand
to make red blood cells in a future age. *Magnesium* from the
sea is another critical element because of its key role in chlorophyll.

Ocean water has *phosphorus* dissolved from igneous (fire-
made) rocks where the phosphorus became linked with minerals
when the earth's crust cooled—this is the key element in the
so-called powerhouse of every living cell (the mitrochondria).

The ocean holds a good supply of *sulfur*, poured into it in flaming lava and by hot sulfur springs such as those that steam among bright-colored rocks in Yellowstone Park—sulfur is concentrated in the hairs of animals, and creates fur, hoofs, and horns. *Silicon*, the element of glass, is dissolved from water-eroded granite, and makes glass shells for diatoms, some of the first plants in the ocean; surely it was a stroke of genius that gave this greatest photosynthesis factory on earth a strong box with transparent roof and walls that let through plenty of radiant sunlight for its operation.

Calcium, the builder of limestone, is everywhere diffused in ocean water. In ancient days it made armor for animals much heavier and stronger than the glass shells of diatoms, and later it gave mineral strength to bones. Calcium armor enabled Lingula (little tongue), a genus of shellfish about an inch long, to survive little changed from Ordovician times (450 million years ago) to this day, to be called by biologists "the oldest animal on earth." Perhaps the greatest contribution to life on earth was atomic calcium in the ocean, which gave animals like snails, barnacles, and crabs shells to survive for a few hours between the tides—a good boost to life's emerging from the sea.

Even rare elements dissolved in the salty sea (which is now considerably saltier than the early oceans)—with comparatively few atoms widely distributed in ocean currents—were somehow collected from far and wide by organisms with inherent zeal to live. *Iodine*, one of a quartet of potent, poisonous elements created in the "white stars" of the cosmos (with fluorine, chlorine, bromine) is scarce in the rocks of the earth's crust, and is so rare in sea water that man doesn't extract it from the ocean. Iodine in the little bottles in drugstores comes from nitrate deposits in Chile, where it has mysteriously accumulated. Yet this stray element on our planet is hoarded by big tough seaweeds. One of these is a kelp off the coast of Chile that grows from the bottom of the sea to the surface in water 266 feet deep, making it the rival of redwood trees as the tallest living thing on earth. Iodine has peculiar potency!

Silver is concentrated in many fish and marine animals. *Cobalt*, with the "least trace" in the ocean, is important in the little cells of the blue-green algae, and it is used in enzymes of various animals and plants. We noted that sea squirts manage to find rare *vanadium* for their body fluids. Only the element *gold* never played a role in organisms—until a recent animal discovered it is fine for filling teeth. But ocean water has everything. (According to Rachel Carson, every cubic mile of ocean water holds about $93 million worth of dissolved gold.)

What a dramatic difference these precisely selected, rare, and often hard-to-find atoms make in the shapes and behavior of living things! A jot here, a tittle there, and the structure of hydrogen, oxygen, carbon, and nitrogen is transmogrified—not too big a word for this astounding result.

The collecting of elements in the ocean, although capricious, resulted in some permanent arrangements—a few basic "blueprints" which became lines of descent. Doubtless many models could not endure the rigid test of natural selection and vanished in the ancient ocean, yet what is left is quite a fantastic throng. Today biologists find about eleven phyla of plants and sixteen phyla of animals, most of them still under water. How amazingly few considering the arithmetic of atoms in the ocean!

In this perspective, all *land* plants and animals (including insects and birds) are rather tentative by-products of ocean life, running dire risks. Only a slight change in the quality of the air or water used in their bodily operations, or a few degrees of the average temperature in which they must live, would wipe them out—while life goes on as usual in the ocean.

Consider how some of the elements in plankton—ceaselessly wandering and drifting and cooking in sunlight through ages and ages—turned up later in the river of life. For example, iodine, an element in blue-green algae (that earliest known form of life on earth) and the rare element hoarded by big brown seaweeds, is vital in the human thyroid gland. Moreover, as we shall see later, in the matter of sex, brown seaweeds appear to be an important forerunner of the human animal.

Trees are an outstanding example of a basic form of life that originated in plankton. Surely nothing is further from ocean life than these biggest of all land organisms. No tree ever grew under water. They are by far the most successful kind of life ever invented for living out of water. The durable, weatherproof bark is impregnated with silicon. Moreover, seeds and fruits are highly charged with phosphorus. That fiery element which is used for match tips is borne in tree sap to buds when they are about to burst open in the spring, and spurs the lengthening twigs. Trees also use sulfur at their root tips, where it stimulates millions of root hairs to bloom; the root hairs (the water-gatherers of the tree) grow so fast you can see them thrust out in a microscope. Tree leaves inherited their chlorophyll from ferns, and ferns from seaweeds. A single atom of magnesium in every chlorophyll molecule promotes the work of the chlorophyll as it splits water and packages sunlight—as it did through the ages in the ocean.

So the atomic medley in sea water—in the course of, say a billion years—produced the dazzling diversity of our living world. A profoundly interesting fact is the orderliness of the hordes of plants and animals of our world. Merely four lines of descent of plants, and five lines of descent of animals, emerged from the ocean to populate the land.*

The well-ordered arrangements made it possible for biologists to detect relationships—how a fly resembles a lobster, how a rhinoceros resembles a mouse, and so on—and to classify and name the species of living things.

This classification indicates that out of infinite numbers of combinations in the ocean only several body plans were good enough to survive out of water. Even these had to have elaborate arrangements for keeping every cell in a pseudo salty sea in body fluids. The flow of dissolved minerals and food, and the carrying away of wastes, had to be maintained just as in plankton.

It is interesting to imagine what life on earth would have been like if the blueprint for that single phylum, the *Chordata* —backbone animals—had not turned up. Without that coinci-

* Plant Phyla: 1. Fungi and lichens. 2. Slime molds. 3. Liverworts and mosses. 4. Ferns, trees, and flowers.
Animal Phyla: 1. *Aschelminthes*—roundworms and nematodes. 2. *Annelida*—earthworms. 3. *Mollusca*—snails and slugs. 4. *Arthropoda*—centipedes, spiders, insects. 5. *Chordata*—reptiles, birds, mammals.

dence, it seems that the Plant Kingdom would have dominated the continents, with tiny creatures like earthworms and insects the chief land animals.

The discovery of the womb of organic evolution in plankton, and seeing it actually working, reveal that *species* of plants and animals of our living world are not an automatic result of cosmic forces. Granted our view is myopic, but judging by the evidence, it took far more than a planet the right size, the right distance from its sun, plus water, to make people. The planet had to have particular kinds of rocks in its crust, with certain proportions of minerals and other elements. Moreover, the atmosphere had to be a mixture of certain gases in precise proportions—including a touch of ozone (a kind of oxygen molecule that is poisonous to life) to screen deadly ultraviolet rays in sunlight. Above all, an immense ocean had to girdle the globe —ever mixing and mixing dissolved electrified particles for such a long time that continents drifted around, sunk under the sea, and arose again to make dry land for another million years or so.

PLANKTON POWER

Why was the plankton world the chief environment on the planet where the foundations for species of living things were laid? Why did no phylum ever originate on land? The answer is rousing and wonderful.

The deliveries into the plankton world of both atomic power in salty sea water and of sunlight energies have continued through the ages. When the populations of one-celled organisms exploded under this influence and plankton formed, this added life power, exerted in the surge and drive of crowds of microscopic organisms. This addition of energy stored in life stuff became a ready-made food supply. Life built on life. Plankton became the greatest concentration of life power on earth. In this milieu, dynamic structures made by one-celled organisms became strong and durable and proceeded to evolve by natural selection.

66

The multitudes of bizarre creatures in plankton today—kinds of life not seen elsewhere in the world—show its terrific creativeness. But this power which produced our living world rests on the surprising paradox that one-celled organisms have not changed much since they first appeared. The minute items at the base of the tree of life—represented by blue-green algae and bacteria—are a mechanism that is never obsolescent. They have never been superseded by anything better for carrying on life. When they multiply, the copies are just as good as the original. Even though cells evolved and acquired odd forms and peculiar equipment, the original one-celled blue-green algae and bacteria were not put out of business. The mass of *bacteria* in the world today is estimated to be twenty times the mass of all animal life on earth. Bacteria reproduce so fast that if they were not checked by a fierce competition for food, the progeny of one bacterium would cover the globe with a layer of bacteria a foot deep in 36 hours.

The tremendous life power of the tiniest organism is no mystery, it is a law of geometry. The *smaller* a thing is, the *greater* is the ratio of its surface to its interior. The interior of a living cell holds vital organs that must be supplied with food dissolved in watery surroundings, and whose wastes must be carried away. A one-celled microscopic creature has its vital organs within microscopic distance of surrounding currents, and enjoys a larger surface in proportion to its size. This gives a tiny one-cell speck of life quicker traffic with its surroundings, and proportionately more traffic, back and forth through its skin —which spells greater life power.

Incredible multitudes of bacteria swarm in the drops of plankton water. They act like chemical laboratories, converting nonliving ions and molecules into nourishment. These are consumed by other one-celled organisms and thus plankton gets the extraordinary life power that drives evolution full tilt.

The nets of oceanographers are scooping plankton from the oceans down to the limit of the penetration of sunlight. "Plankton Puzzle No. 1" is the discovery that the animals in this weird world are not just drifting passively—except as a whole mass is carried on worldwide tours. Individuals jiggle their bodies or vibrate odd projections to swim vertically—going up at night, and down in the daytime. Tiny creatures go hundreds of feet in an hour—never sidewise, the ocean currents take care of that motion. The champion can dive in short two-minute bursts at a speed of up to 700 feet per hour. This plankton animal, named *Meganyctiphanes* (big-night-torch-carrier), has spots that glow red in the dark, four on the abdomen, and a luminous red spot in each eye.

Each plankton animal has its own peculiar method for going up and down, moves at its own speed, and seeks its personal depth. Thus the plankton animals are distributed vertically in the water, and each one has its share of living space. So plankton never becomes a dense, sticky mass of organic stuff; it is dynamic with life power, free-moving and fluid.

Many theories have been advanced for this strange behavior of plankton: the animals going down in daylight, when their food, the green plant cells, are denser on the surface since plants turn toward the sunlight. Alister Hardy (see page 118), who has delved into this riddle, supposes that vertical migration gives plankton animals a continual change of environment. Surface currents move faster than deeper currents, and their meandering provides ever fresh pastures.

THE POWERHOUSE OF WATER AND LIFE

That seething, bustling plankton world must have some high-powered green plants to give it food energy, as do all realms of life on earth. What kind of plants package sunlight for those fantastic hordes in the ocean? The familiar seaweeds that drape coastlines are localized, and offer fine hiding places for fish and lobsters and crabs in the tidal currents of the continental shelves. Sargasso weed makes a huge whirlpool in mid-Atlantic, a standing crop of 7 million tons of seaweeds. That is a myth about a labyrinth that entangles ships and threatens ancient mariners with monsters. The Sargasso Sea has loose clumps of weeds that are sometimes hard to find, its surface is dark blue, and there seems to be less plankton life in it than in any other part of the worldwide oceans. Compared to other ocean areas it is a "biological desert."

So what plants drive that ancient engine of evolution, plankton —which is the same kind of operation in all the oceans, the North and South Atlantic, the Pacific, the Indian Ocean, and sea water between and under the ice of the Polar regions.

The answer is, the microscopic plant named *diatom,* which means "cut-in-two." Wherever you see ocean, this unbelievable plant is spreading its "green leaves" under the surface and tumbling in the surf. It is estimated that there are enough diatoms for one in every drop of sunlit sea water. Diatoms multiply so fast that they can double their population in a day.

So why don't diatoms choke the oceans with a solid mass of their bodies? Because these packers of massive sunlight energy are the basic food for multitudes evolving in their midst. Diatoms are devoured as fast as they multiply. We have noted how the plankton beings were organized in phyla—and how some of these in a recent age emerged from the sea to populate the land. So diatoms, at the base of the pyramid of all life on the planet, are "the most important plants on earth."

What about photosynthesis of forests, grasslands, and grain fields? The oceans spread a many times greater light-catching surface around the globe than do the land areas—with an infinity of diatoms trapping even dim rays of light to a depth of some 50 feet. Adding to diatoms the many other kinds of ocean organisms with chlorophyll, and the big visible seaweeds—scientists calculate that some 75 percent of all photosynthesis on earth takes place in the ocean.

Little "cut-in-two" is not only the most important plant on earth, it has also been called the most beautiful. (That is relative, and personally I would vote for a maple tree, not counting wildflowers in the competition.) People who look at a diatom through a microscope see a glass box, which may be circular, square, shield-shaped, oval, rectangular—there are many species of these plants, each with its characteristic shape, but all of them are glass boxes with one half fitting over the other half like the overlapping lid of a candy box. Each of these extraordinary specks of plant life is finely etched with geometric lines and circles and fountains. Each species of diatom has its personal etching pattern, as we have our personal fingerprint swirls. Seen through electron microscopes, the wonder of their exquisite etchings grows. The pure glass diatom box is filigreed with such fineness that a human hair would have to be sliced lengthwise into four hundred slices to fit between the marks.

Diatoms multiply by cell division, like most cells. But how does one divide a glass box? There is only one way to do this without broken glass, and that is to lift off the lid. This a diatom does with great frequency—each part of the box then goes off on its own and swiftly makes itself a new glass half that always fits *inside* of the old one. This leaves one progeny that was the outside lid full size, but the other progeny that was the part of the box fitted inside the lid has to be a *smaller* diatom.

As the clouds of diatoms multiply, 50 percent of them will hold the natural genetic size of the species, the other 50 percent will steadily become smaller and smaller. This peculiar threat to the existence of diatoms is met by an episode that adds to the fairytale tinge of the plankton world. For some generations the littler ones carry on, but when they reach a size that is "too small for any good use," the undersized diatom stops dividing by lifting its lid. It now throws away its glass box, and the soft shapeless body holding the precious chlorophyll floats naked in the salt water while it grows to full diatom size and makes a new glass box.

The accumulation of glass boxes on the ocean floor from this episode—and from diatoms that are eaten whose glass, expelled as indigestible, sinks to the bottom—gives an idea of the magnitude of this life power in plankton. As the glass snowstorm in the ocean continued steadily through millions of years the currents gathered deep drifts of diatom glass. Richmond, Virginia, is built on an ancient ocean floor with such a drift, where diatom glass is 40 feet deep. At Lompoc, California, a drift of diatom glass is 700 feet deep in places, and is mined for human use—as an abrasive called diatomaceous earth.

The whole drama of water and life can be seen in sharp focus in this glittering little green plant. It represents a peak of evolution. Diatoms never evolved into anything else or anything bigger. The plants of our world—ferns, trees, wildflowers, grasses —did not come from the diatom line, they came from the big seaweeds which attached themselves to rocks on the coast by holdfasts, and stalks, that led to the evolution of roots, stems, and green leaves.

So diatoms are a dazzling "dead end" of evolution. They didn't need to evolve into anything different to be the most successful plants on earth. See their beautiful equilibrium in the interplay of water and sunlight. They turn this way and that, catching light from every angle as they dance, extremely buoyant with light finespun porous glass, and with droplets of oil in their bodies. The transparent glass box freely transmits radiant light energy. What is more, the etching patterns are deemed to act like prisms that focus light rays on the chlorophyll molecules.

Thus the filigree of the glass, some of it so fine it took the electron microscope to discover, has peculiar significance. It enables diatoms to operate in deeper, darker water, and cloudy days do not materially reduce the food-making powers of this marvel.

The floor of the ocean is, in a sense, the strangest "planet" being explored today with space age techniques. I refer to the abyssal depths which lie beyond the continental shelf. The shelf extends various distances from shores—about 10 miles off California, 60 miles off New York, over a hundred miles on the Grand Banks off Newfoundland. At the shoulders of the continents, steep slopes plunge into blackness. Off the mouths of big rivers fabulous canyons cut through the rim of continental blocks. For example, a canyon 60 miles off the mouth of the Hudson River is greater than the Grand Canyon. Two miles down at the foot of the Hudson Gorge there is a vast flat plain that stretches out under the mid-Atlantic. A pinnacle of rock rises twice as high as the Empire State Building exactly at the bottom of the Hudson Gorge, like a colossal monument—never to be seen by tourists—at the entrance to the Atlantic Abyssal Plain. The flat abyssal plains of all the oceans cover more than half of the earth's surface.

Oceanographers are excited about the stores of mineral wealth down there, but we are interested in the carpets spread by fall-out from the plankton world miles above the abyssal plains. Here is astonishing evidence of the life power of plankton evolving our familiar world of plants and animals.

We have noted the massive drifts of diatom glass piled up by ocean currents on continental shelves. But out beyond, miles under the blue surface, undisturbed by wind and tides, ocean currents creep only by the slow mixing of colder and warmer water that smooths and flattens sediments like a giant, silent steamroller. In this way, mysterious carpets have been laid through millions of years on the bottom of the seas.

These carpets on the ocean floors have a simple technical scientific name—they are *oozes*. Ocean floor ooze is not sticky and muddy, and it is startlingly different from ordinary sand and clay sediments. It is quite firm and clean, and when hauled up into the daylight, oozes reflect beautiful colors, according to where they came from—they are mostly cast-off particles from animals in the plankton floating miles above.

If the ocean floor carpet comes chiefly from the remains of *Globigerina* (the globe-carrier) the ooze will be limy, made from calcium, and its color will be milky white, rose, yellow, or brown. This is the floor covering under two-thirds of the Atlantic Ocean.

If it comes from the remains of *Radiolaria* (the radiant animals) the carpet is red, chocolate brown, or straw colored. This is the floor covering chiefly of the South Pacific and Indian oceans, in water more than three miles deep. Diatom oozes also cover large areas of the ocean floor, particularly in cold waters of the Arctic regions.

Considering that these plankton animals are one-celled and microscopic, the infinite numbers of them are beyond comprehension—their cast-off skeletons carpet 60 percent of the ocean floor.

The depths of these amazing floor coverings are beyond calculations, but judging by geologic features like the chalk cliffs of Dover, which were deposited by microscopic creatures, they may be hundreds of feet deep in places. They tell the story of the evolution of life in plankton through millions of years. It is estimated that, on the average, oozes deepen at the rate of $\frac{1}{25}$ inch per hundred years.

The action is still going on, exactly as it has for millions of years. The ocean scene of evolution is stable. Of course, evolution in plankton is proceeding without a pause today. The reason we do not see the creatures change or new lines of descent turn up is only because of the tiny span of our human moment on earth.

7

Nothing Else Like This in the Solar System

Just imagine what a field day the TV commentators will have if any kind of life is ever discovered somewhere other than on planet earth. Our space scientists tried hard with earnest conjectures about microbes on the moon—and just in case, returning moon-walkers were rigorously quarantined. The laboratories made a yearning, meticulous search in the moon rocks for a hint of an organic molecule, a carbon compound, or a trace of sedimentary rock formed by water. They scrutinized the rock molecules with electron microscopes, and linked spectrographs to computers.

The reports make geologic history. They contain phrases such as "impact-fused glass beads" (the impacts of countless meteor collisions with the moon) and "vuggy shocked breccia" (that is to say, mineral gravel shot through with crystals). The moon rocks were distributed among 142 scientists in separate laboratories, and their findings are summed up by Dr. Brian J. Skinner, Chairman of the Geology Department, Yale University, with these words: "We all found the same things . . . the moon samples show about as unappetizing a diet for any form of life you can possibly imagine."

Thereupon, the bugle call—To Mars in the 1970's!—summoned the astronauts for another try. What would life on Mars or anywhere else in our solar system look like?

74

Perhaps it has no more shape than chemical droplets, like the microspheres mentioned in Chapter 5, or it may be an extraordinary molecule as magnificent as protein or chlorophyll. A popular speculation is that of little beings the size of insects, which may have achieved an advanced civilization with electronic skills.

It has been suggested by biologists that plant life on Mars will be like lichens on earth; these can exist on the sun-scorched face of a boulder in a desert, and they are also unfazed by temperatures deep below zero in the polar night. Another scientific surmise is that life on our sister planet may be akin to blue-green algae or cocci bacteria which can live in boiling hot sulfur springs (as seen in Yellowstone Park), and in petroleum wells underground, and in muck at the bottom of a bog cut off from sunlight and oxygen.

Note that these speculations all assume that there is water for their creation, maintenance, and reproduction. But the towering fallacy is to regard them as crude, elemental life.

Lichens are modern pop abstract art with two motifs—either fluid, lobed discs, or eccentric little trees often tipped with bright red beads. A lichen has no DNA code of its own, no descendants. It is created eccentrically by the merger of a number of single-cell algae and a fungus. The individual they create bears no resemblance to either of its constituents. This is a dramatically successful design for living, where the algae, with chlorophyll, supply food for the corporate body, and the fungus, with a wad of cottony fibers sealed inside a durable, impervious crust, holds body fluid secure from evaporation.

And lichens are not so humble as they look. They brew a powerful acid that dissolves rock crystals fast. Limestone, and even garnets, mica, and extremely hard silicate crystals in granite melt at the touch of the lichen, which then thrusts threads into the opening and becomes as permanent as the rock. It is these acids, unique chemicals in nature, that give lichens a silvery sparkle. Their salts are sensitive to light; on a bright, clear day lichens change from silver to yellow to sealing wax red. They are certainly not raw and crude—they are entrepreneurs in air, water, and sunlight.

Blue-green algae and cocci bacteria are just as far-fetched a speculation about life on another planet. They are not crude, elemental life able to sit in a hopeless fix in a vicious situation. They are elaborate living cells which, in the words of a distinguished microbiologist, "exhibit profound resemblances to man, in their physical makeup, in their chemical activities, and in their responses to various stimuli." The electron microscope perceives exquisite, delicate structures. Inherently they are as fine life as we are. Sexuality has even been detected among bacteria. The chief difference between us is size, and the fact that they didn't seem to have to undergo dramatic evolution to survive. Microbes have a more legitimate title to planet earth than do we. The total weight and volume of microbial life is some twenty times greater than that of the total animal life.

Scientists have theorized that liquid ammonia might summon up some kind of life where there is no H_2O. In addition to performing as a liquid it offers an energetic combination of hydrogen and nitrogen. The hitch is that ammonia is liquid only in a very narrow range of temperature—and if it is to be permanently liquid the temperature must not get higher than 100° below zero. But suppose there is a planet somewhere with a climate like that?

It is estimated that the processes that led to the origin of life on earth—the first living cell—took a billion years, with water doing the work of assembling the raw elements. Biochemists figure that, considering the way cold slows down chemical reactions, it would take liquid ammonia 64 billion years to do the same job. This is far greater than the age of our galaxy, and far greater than a star like our sun could supply radiant energy.*

This brings us splashing back to earth. Instead of trying to stare at Martians through very high power microscopes, let us take another look at the plankton world, to find out how our living world shaped up. The fantasy of this "other world" that is not out there but right here on earth with us, is well described in the words of a famous poet:

---

* George Wald, *The Origins of Life*, Proceedings of the National Academy of Sciences (August, 1964).

Curious, eager wits pursue
Strange devices weird and new,
Like the scene you witness here,
Unaccountable and queer.
I myself, if merely told it,
And I did not here behold it,
Should have deemed it utter folly,
Craziness and nonsense wholly.
—ARISTOPHANES, 400 B.C.

This does not imply that these "queer" kinds of life are out-moded antiques which have been superseded by the conspicuous plants and animals of our world. To the contrary. They are masterpieces of evolution, triumphs of natural selection, superbly fit to live in the original, most permanent, and best place to live on earth—the blue salty sea.

The lively creatures in the ocean show how all sorts of body shapes were originally tested—those in plankton today are the most successful types, according to the law of survival by natural selection. Here are various appendages which foreshadow limbs, fins, and wings. They reveal how nerve systems developed from mere sensitivity in quivering protoplasm to exquisite organs for touching, tasting, smelling, seeing, and hearing.

These strange multitudes are also perfecting a revolutionary way to multiply. The time-honored method, which started with the first living cells, was simply for one cell to divide into two portions of protoplasm. In the plankton world some beings are dallying with reproducing by becoming two kinds of individuals, male and female. This was destined to have a tremendous impact on future life on planet earth.

The plankton world clearly shows that not every organism necessarily evolves. Selection depending on chance mutation in a genetic code is a dizzy game of chance. In plankton we look behind the scenes to see the exciting roulette in ferocious opera-tion, spurred on by the electronics of salty ocean water.

Here are the primordial organisms, simple algae and bacteria, still in their original forms. Together with them are one-celled organisms which are innovations, such as the amebas. Then on up the ladder of life, they are all here on their rungs, to fish, sharks and whales, which are modern monsters—they are not plankton members, but they couldn't exist unless they invaded it to gulp huge mouthfuls of plankton organisms for food.

Ameba is studied in first year biology as a lowly, simple creature. Viewed through an ordinary microscope, the speck of colloid seems hardly more than a droplet of watery jelly, yet it is sensitive and moves around in a deliberate way with an air of dignity. It pokes out a bulge called a *pseudopodium* (false foot). Then the body contents flow into the foot, and thus it rolls over rough ground like a caterpillar tractor—ameba is so tiny that all ground is rough to it. Since the false foot can poke out from all over the body, ameba can head in any direction without turning, and reach in all directions for food, mostly diatoms and bacteria, which it wolfs by engulfment. After digestion, aided by enzymes pumped into the little pocket that holds ameba's victim, wastes leave the body the same way by reverse sorption through any part of the body surface.

This blob of jelly fits its water world perfectly. In the way it prowls for food, senses good things to eat, and, when it finds them, pokes out its "false foot" to touch and absorb the food, it represents a superb leap in evolution compared to passive blue-green algae and bacteria. These eaters of algae and bacteria might have been the dominant life on earth for all time, if bigger creatures had not turned up and devoured amebas in their turn.

Let's see what happened when the timeless ocean went to work on them. Some of ameba's relatives developed a sense for detecting calcium atoms dissolved in the water, others a sense for detecting silicon atoms. The former became Globigerina and the latter the Radiolaria. We noted in the last chapter their massive achievements in the way they paved the ocean floors with wet cement and glass oozes. The superb globe animals ignore algae and bacteria as food, but feast on the diatoms which teem in the ocean.

The salt water grows ever richer with the food of dead bodies, thus spurring the evolution of ever more grandiose organisms. The globe creatures spin spiral shells with calcium—a portent of colorful snails to come. Moreover, the globe animal can swim around a bit by waving threads, and it has puffed out its body to explore a greater area of water for calcium and food. Swaying in the water, the living threads stick together when they touch, forming a kind of spider web in which all sorts of particles of food are entangled. This food it digested *outside* the body in the sea water by the living web, and the nourishing juices flow through the threads into the spiral of globes that make up the body of a globigerina. Thus, the globe animal neatly taps the ocean currents, focuses its energies on its reproducing organs, and achieves incredible multiplying.

Radiolaria, which collect silicon atoms, show how natural selection makes different structures under different conditions, for food-getting and reproducing. In the radiant animals, silicon is combined with oxygen. (This is the same chemical combination with which man makes glass.) Long, sharp glass needles poke out through the skin of Radiolaria and project in all directions—a bristling contrast to the globe animal's flexing fishing net. The glass needle is coated with protoplasm, which makes it sticky all over, and can make it grow longer by adding glass to the point.

With needles radiating in all directions, moving and groping like the spines of a sea urchin, the food catch for such a tiny single-celled animal is great. The food particles flow in the protoplasm coatings of the needles into the central organs of the creature, where digesting and reproducing are in gear together. This mechanism, continuously fed energy by the currents, makes the radiant animals explode with vitality and multiply as fast as their cousins, the globe animals.

Another of the plankton world's marvels is a dazzling example of innovations which occurred in creatures with only one cell for a body. People on cruises in tropical waters crowd the rail to see "phosphorescence" in the ocean waves. The spectacle, appearing as though the sea water itself were afire, has long stirred awe and wonder. Charles Darwin describes such a scene in his book, *The Voyage of the Beagle*:

On a dark night off the east coast of South America the sea presented a wonderful spectacle. There was a fresh breeze and the foamy surface glowed with a pale light. The vessel drove before her bow two billows of liquid phosphorus, and in her wake she was followed by a milky train. As far as the eye could see the crest of every wave was bright.

Even Darwin did not recognize that this is a display of live creatures. *Noctiluca* (night-light) would be known only to people with microscopes, except that it is one of the little animals that makes ocean waves sparkle on a dark night. Single-celled night-light imitates a rubber balloon filled with a fluid lighter than water, which makes it buoyant so it bounces around near the surface. The diatoms, needing light for photosynthesis, accumulate in greater numbers near the surface, and these are the chief food of *Noctilucas*. So, by being buoyant, the creatures stake out their claim to green pastures. Moreover, surface water is windblown this way and that, which gives them strenuous mobility for foraging without exerting any energy.

Evolution both develops devices that promote living, and downgrades equipment that is unused. Night-light is a type of microscopic animal called a *flagellate* (little whip), characterized by two long strong hairs which are waved furiously to be on the go. In this case, night-light merely jerks *one* of its whips slightly about every six seconds, and uses the other hair to rotate feebly.

The flashes of light occur in tiny strings of beads under its skin and draped through its body fluid. When they light up, the animal is a scopic transparent ball festooned with strings of sparkling lights. The creature's mouth is a dimple, like that at the top of an apple. The flashing starts in a circle of beads around this dimple and spreads in waves all over and through the body.

There is no evidence that these flashing lights are signals like the mating flashes of fireflies, or to scare off predators. The brightness seems to happen when night-light is dealt blows by whitecaps or in the wake of a ship. The phosphorus animals do not light up in calm water—they only light up when agitated. Those strings of flashing beads are precursors of nervous systems in an age when one-celled animals were the inhabitants of the planet.

So the plankton world became ever more dynamic with vital energies, with the enormous production of packaged sunlight by ever greater multitudes of diatoms, and from the perishing of hordes of bacteria, amebas, globe animals, radiant animals, night-lights, and countless others.

Why, if conditions in the ocean were so ideal for life to flourish in the one-cell size, did not microscopic beings in the salty seas remain the only kind of life on a planet like ours? What happened to change the picture?

Two dynamic phenomena were set in motion. First, organic evolution in salt water accelerated until it "boiled over!" Second, the atmosphere of the juvenile planet became charged with more and more oxygen from burgeoning populations of diatoms, algae, and photosynthesizing bacteria which throw away oxygen atoms into the water.

The first event caused one-celled organisms to grow to the limit of their size and elaborateness, until some stuck together, creating *metazoa*, the many-celled animals. Thereupon, fired by more oxygen and light, the forces of evolution gained ever greater momentum, the fight to survive became ever more ferocious among plankton giants—until a few of them were driven out of the ocean and staggered across the beach, desperately trying to survive.

The whole vast spectrum of life as it has evolved to this time still depends on the ancient one-celled hordes. The free oxygen in the air would be all used up in 2,000 years at the present rate of breathing were it not for diatoms in the ocean, the chief maintainers of breathable air. Moreover, bacteria in our bodies are indispensable to our existence—they attack other microorganisms that would destroy us, they supply vitamins, and the bacteria of decay make fertile soil that produces our food, flowers, and trees.

THE FIRST GIANTS

Those single-cellers were, to be sure, a towering success as a way of life. They could rise and sink by expanding and contracting their bubbly little bodies, they could move an inch or so by waving hairs, but a radical new invention was needed the instant they welded together and made bigger and heavier structures. Animals that brandish tentacles, that move by jointed limbs of free-swimming tails or fins, must have muscles.

The contracting muscle fiber, which can lengthen and shorten powerfully, seems to have been invented only once, and after that it was taken over by the Animal Kingdom. All kinds of creatures use the same basic principle of contracting muscle fibers—jellyfish, oysters, starfish, snails; joint-leg things such as crabs, lobsters, insects; bony animals such as fish, snakes, birds, mammals.

We tend to take muscles for granted, yet this astounding mechanism marked a turning point in organic evolution, and led to races of giants, which could swallow billions of one-cellers at one gulp.

Muscle fiber is made of two common proteins (myosin and actin) arranged in two layers that slide back and forth over each other. This action is instant and lightning fast, and mostly under the control of its individual. In primitive metazoans it responds to a subtle sensitiveness of protoplasm. In bigger bodies, nerve systems evolved that supplement instincts in operating muscles.

Take a look at a giant, sluggish metazoan without muscles. The sponge couldn't evolve as a respectable, active animal without muscles. People who go after sponges at Tarpon Springs, Florida, with long-handled pruning shears think they are some kind of seaweed. However, this is a real animal, although it has no head, no appendage for gesticulating, no organs or nerves, or particular shape, and its body fluid is merely the salt sea water that flows through its canals whose openings are seen as the holes in a sponge.

The sponge is made of astronomical numbers of amebas and hair-wavers (flagellates) stuck together to compose an extraordinary sort of skin that forms the tubes of the sponge body. The giant has tough elastic skeleton, as you know when you squeeze a sponge, and the body of a living sponge has jelly between in which amebas travel far to do odd jobs in building the animal and adding their own bodies as building material. As a living jelly sandwich it can grow to any size. It is a masterpiece of one-cellers before they surrendered all individuality to make organs and structures. Any tiny fragment of a live sponge can create a full-size sponge.

This least animallike animal is fixed to a spot and it has no coordinating nervous system. The brainless thing, for a while, grows by cell division to any size, with all sorts of shapes and patterns.

It is hard to imagine a truer ocean animal than a jellyfish. If one is cast upon a beach in the surf and dries in the sun, it vanishes—it has evaporated.

A jellyfish's body is 99 percent water with no brain or central nervous system. Yet its body is a delicately sensitive tissue. Stimuli are received by surface cells all over the body and picked up from them by a nerve network that forms a ring around the rim of the bell-shaped body. This transmits information to muscle cells which contract and expand the bell and to sex glands which secrete eggs or sperm. R. W. Moncrieff * says that this brainless animal "by its reflex actions gives the appearance of experiencing pleasure or displeasure." This important human attribute had to start somewhere, why not in a jellyfish?

The jellyfish style of body reached its summit of evolution with Portuguese man-of-war, an eerie and fearsome monster of the plankton community. The body dome is a foot across, but the animal hunts in the depths of the plankton with swaying tentacles some 30 feet long, which form a curtain that drops from the rim of the bell. The bell does not pulsate to give mobility. Instead the monster has developed a lofty crest which it

---

* *The Chemical Senses* (New York: John Wiley & Sons, Inc., 1946).

erects to catch the wind. This can be turned diagonally, to sail upwind like a sloop. The crest is filled with oxygen-nitrogen gas that, caught in the sunlight as it sails along, flashes with iridescent blue.

The loveliness of the monster as a colorful sailboat is all above water. Under the bell countless prey are being stung by poison arrows shot from the tentacles, which can paralyze even large fish—and bathers flee from the surf when a Portuguese man-of-war is sighted.

This creature exhibits the amazing way the melee in the plankton world tailors creatures to fit into all sorts of niches and opportunities for food. Those long, deadly tentacles enclose an area that is safe from predators, while also holding much animal and plant food near the big mouth of the gluttonous Portuguese man-of-war.

So a certain clever little fish appropriated this refuge behind the circular curtain of the tentacles as a homeland—*Nomeus grovonii*, or "cavern pastures." The cavern of this little fish is the area enclosed and cut off from the rest of the sea behind the curtain of stinging tentacles of the Portuguese man-of-war. Their pastures are the many organisms swept up by those tentacles with some nourishing solutions from the dissolved parts of big prey.

This fish is precisely the right size to use this niche in the manner of goldfish in a bowl. They go wherever their niche goes, while no hungry predator dares to touch them. Those tentacles paralyze large fish, and fishermen are in awe of them —if an arm is touched by a tentacle entwined in a tow net, it gets an agonizing sting. The deadly drapery, which trails out as the Portuguese man-of-war trawls the sea, is bright blue. The little fish's vertical, iridescent blue stripes on silver bodies makes them invisible under the blue dome.

This prompts Sir Alister Hardy to comment on the engaging mystery: "It seems to be a remarkable case of mimicry—for the intense blue tentacles must be avoided like the plague by predatory fish. Are these little fish immune to the paralyzing sting, or do they deftly avoid touching the tentacles? Are they engaged in some subtle partnership with their overlord? It is one of the many puzzles of pelagic natural history yet unsolved."

The little fish taking transatlantic trips in utter security and luxury is a beguiling illustration of a species designed "by Darwin" to fit a precious situation where everything, even the guns of a fortress, insure its survival.

SEA SQUIRT'S GHOST STORY

Ciona (Greek for "demigoddess"), the sea squirt, reveals a sensational leap in evolution. The adult is a brown bag about 4 inches long with irregular knobs and bumps, about as comely as a potato. It has two holes at the top end; one takes in water, and the other is a vent that squirts it out, enabling sea squirt to dart around by recoil, the same principle as that of a jet airplane.

Most of the body is an immense pharynx that forms a sac extending almost to the base, where the animal's stomach, intestines, heart, and reproductive organs are crowded into a small space. This body plan insures a great food supply. The fat glutton sucks in nourishment and makes a thick soup which it stores up in the huge pharynx. Such a design for living insures success, especially since the animal can drive off predators with forceful squirts. Many sea squirts are flourishing in today's ocean.

What makes sea squirt of special interest is that, in a sense, it is an ancestor of the human race. It may reproduce in the primitive way by budding, and each bud will grow into a new sea squirt; on the other hand, the beast can also produce eggs. When an egg hatches, a little wiggling tadpole emerges. This tadpole has a tough cord, called *notochord*, running down its back along with a nerve cord. Biologists consider this tadpole form of sea squirt as an ancestor of the Chordata—the backbone animals.

One other feature of the unbelievable sea squirt is cellulose—the chief substance of wood. A tree is an emblem of true land life largely because of wood, the ideal material for stiff trunks. Seaweeds do not have any wood, and no member of the animal

85

kingdom has any wood—at least, that was the opinion of biologists until recently when it was discovered that there is a trace of cellulose in human skin. Sea squirt's tough, leathery skin is rigid with cellulose! The ancient, ugly sea squirt seems to have been a kind of forerunner of both people and trees.

8

From Blue-Green Algae to Man's Brain

Two colossi loom up among the buoyant plankton beings—of a huge new order of magnitude, they look and act very differently than "those who wander and drift." They are the *Arthropods* (joint-leg animals) and the *Chordates* (animals with cords in their backs), destined to be the trailblazers upon the land.

They appear quite suddenly in the fossil record, but it had taken several billion years for these pioneer explorers to develop on continents. Crucial problems had to be solved, such as muscle operation, resistance to crushing by gravity, and oxygen intake in volume.

In the course of timeless ocean time, bristles, whipping hairs, and ameba's false feet were succeeded by jointed legs which, although awkward as a staggering crab, were able to walk on firm ground. Some of these lords of creation had developed swim bladders which made their heavy bodies more buoyant— only a step from lungs. And the cords in the backs of the sea squirt larvae were converted to backbones and sinews to work muscles.

But by far the most important feature of the magnificent Arthropods and Chordates was their possession of finely developed senses. The first ones out of the water could see sunlight and shadows, feel warmth and coolness, employ taste and smell, detect pebbles from what is edible—in fact, they had all the senses needed to escape danger and to find food and mates. All five senses were perfected in the primordial sea, and basic designs for sense receptors had been thoroughly tested in the ancient plankton world.

The blue-green algae pioneered photosynthesis with a kind of "chemical sight." They had pigments, sensitive to light—pigments which are the same kind of protein that makes the pigment cells in the retinas of eyes of fish, frogs, snakes, birds, and people today.* The arrangements of these pigments which detect various wavelengths (the colors) in the sun's spectrum is that of "oriented proteins piled one on top of another like a stack of plates." This light-sensitive receptor is similar in blue-green algae, green leaves, and our eyes.

The blue-greens are amazingly sensitive to various wavelengths of light. The blue-green pigment which gives them their name masks other pigments sensitive to other light waves. Blue-green algae also show red, yellow, and brown, and they have carotene, the orange-red pigment of carrots. Blue-green algae change colors dramatically when illuminated with colored lights. A blue-green called *oscillatoria* (it oscillates) turns yellow in blue light, blue in yellow light, green in red light!

In this way, the 2.6 billion-year-old blue-greens were making bold experiments with light waves. A mechanism for survival was set up ages before there was anything like a brain to sort the light waves and to see colors. The sense of perceiving the visible wavelengths of light was inherent in the play of colors in the first blue-green algae.

The evidence of what happened more than two billion years ago is still there to be seen today—how the blue-greens, perceiving the various wavelengths of light, shifted to that source of energy, from the energies of iron and sulfur and phosphorus dissolved from rocks; how they prospered so mightily that there

---

* The evolution of the sense of sight in higher animals resulted from the perfection of a wonderful pigment called rhodopsin (a rosy pigment, reflected in the name of rhododendrons). This pigment, which is found in primitive algae, is also the basic and most influential pigment in the eyes of many animals today. It accumulates in quantity in the rods of human eyes.

88

occurred a leap up the evolution ladder with the creation of another species, the green algae, with which the familiar plant kingdom, distinguished by green chlorophyll, was founded.

This "seeing light" by the primary algae, and by green leaves today, looks like a simple matter—just sunlight striking a green pigment. To the contrary, it poses one of the greatest challenges to the life sciences. Molecular biologists probing the atomic universe inside a chlorophyll molecule, with such techniques as listening to the resonance of spinning electrons, are staggered by the way those primordial organisms selected particular colors and used their energies with utter precision.

Chlorophyll is green because it does not use the wavelengths in green light—but throws them away, reflects them for our eyes to see. Red and blue light waves operate the food factory, assisted by some violet light absorbed by the orange-red carotene pigments.

The laboratories discern sunlight piercing labyrinths of atoms in quick flashes, each $1/100,000$ second, and then a period of darkness that is two thousand times longer—$1/50$ second. This creates a cascade of reactions in which electrons are knocked around and become involved with more than one nucleus, and H_2O's are split apart and their O's set free and thrown off into the air. This chaos of sunlight energies is instantly disciplined and channeled through two cycles, whose interplay performs the miracle of transforming *light energy* into *chemical energy*, namely, food.

The first blue-greens doubtless used the food they made in this way in their bodies, for their own sustenance. However, in the long run, their keen sense of light produced an energy surplus to share with the whole living world. Measured in calories, the energy extracted from sunlight by water in chlorophyll, is about 100 times greater than that generated by an electric plant using the same amount of water in a 333-foot waterfall.*

It is an engaging thought that the blue-greens' sense of light made them so efficient that they never evolved as anything different. The law of natural selection selected them once and for

* J. Terrien, G. Truffant, and J. Carles, *Light, Vegetation, and Chlorophyll* (New York: Philosophical Library, 1957).

all time, as long as there is an ocean upon earth. When you think of the hordes of extinct and vanishing species, this is quite a distinction.

These antiques, together with their green algae relatives, populate our modern world spectacularly with their displays of colors around hot springs. In those locations the water holds strong solutions of sulfur, calcium, and iron—making an environment like that in the elemental solutions of the primordial globe. Astronomical numbers of blue-greens live in the Red Sea bordered by the mineral glare of the Nubian and Arabian deserts. Here they choose somewhat different colors than those used in regular photosynthesis. Solutions of silicon and sulfur from the desert sands alter the spectrum of light in the water, so the algae absorb more blue and green light, and reflect more of the hot red radiance, which is what makes the Red Sea red.

Their sense of light is so fine, and they employ its energies so efficiently, that they can flourish even in darkness for a while, if there is moisture. Occasional plowing lets in enough light and moisture for their breeding. Agriculturists studying soil fertility estimate that an acre of topsoil of good farmland in a plow depth of merely six inches contains, on the average, one hundred pounds of "weightless" invisibly microscopic algae!

In the same way that perception of the wavelengths of sunlight (colors) suggests *sight*, the "feel" of water by blue-greens suggests *touch*.

Each blue-green alga cell wraps itself in a protective gelatinous coating. This may help it to focus more light rays by causing them to bend inward by refraction, and also the sticky coat may serve as a collector of needed molecules from the surrounding water. An interesting result of this asset is that blue-green cells stick together in long threads, while retaining their status as single cells. A spectacle which has long fascinated biologists peering at the beginnings of life through microscopes is the way clusters of these blue-green threads move spontaneously. Their tops coil like question marks and then straighten. They move backward and forward. Often they sway in unison, with ballet rhythm.

What is the sense of it? They have no muscles, the cells have no detectable way of communicating with each other. Molecular scientists, staring at this wonderful performance of the most primitive kind of life, find no explanation. One biologist is quite frank about it—"The mechanism for movement in blue-green algae is not understood."

Millions of years after blue-greens felt the gentle touch of water, a farmer will moisten his finger and hold it up to predict the weather by the touch of the air. And guided by the touch of air, birds will float and wheel in buoyant flight.

THE SENSES OF OCEAN GIANTS

A creature is a giant only in relation to something much smaller; in that way an ameba, one-hundredth of an inch long, is a giant compared to a blue-green alga. This shapeless animal, so long regarded as the lowliest, least complicated thing in the protozoan menagerie, solves problems of food-getting and reproduction with astonishing sensitivity.

An ameba never moves capriciously, but always with an air of dignity, as one accustomed to making the right decisions. With no front or rear it moves in any direction instantly. With no mouth, no stomach, no cilia to wash in food, it swallows and digests. Without reproductive organs it reproduces smoothly by simply splitting in two, each half growing to full size and going its way. Without visible sense receptors this droplet of protoplasm somehow receives signals from its surroundings and makes choices. If you put a speck of red ink beside it with a medicine dropper, ameba recoils with a jerk. Ameba's normal gait when prowling for bacteria, its favorite food, is about 1/25 inch per hour—that is, it travels a distance equal to the length of its body in two minutes. If you lay a thread of agar bristling with bacteria before it, this footless creature "quickens its footsteps," hastens four times faster along the bacteria trail, at a rate of 1/6 inch per hour, feasting as it goes.

The sense of *touch* has been called the earliest of the five senses largely because biology students looking through optical microscopes are entertained watching protozoa flinch at the touch of an unacceptable item, and resolutely detour around a grain of sand. However, pursuit of prey actually indicates a sense of *smell*.

All five senses are similar phenomena. We well know that *taste* and *odor* are closely related chemical senses. Is not the sense of light we call sight the same as feeling the touch of light waves? By the same token, hearing might be thought of as feeling the vibration of sound waves.

The five senses began with the first living cell that appeared in Oparin's soup.

AT LAST—THE APE BRAIN

Geologic evidence shows that what we are pleased to call our superior brain began to shape up about two million years ago—just yesterday in geologic time—when an ape named *Australopithecus* was running around in open country on his two hind legs. By general agreement among authorities, this particular animal is the earliest known ape ancestor of the human race.

In a brutish world, when other apes stuck to home territories, Australopithecus traveled far and wide. Traces of this ape have been found in colonies from southeast Asia and Africa across the Near East and Europe as far west as Spain. What is more, Australopithecus shaped crude tools out of wood and stone, and his towering triumph was to bring fire under control. Apparently he made campfires when fire was a terrifying nightmare to wild beasts.

The theory is that one of these apes was struck by a mutation (a sudden change in an inheritance code) that slightly improved his brain. The advantage made him a little smarter in finding food; this preserved the mutation, and then by crossbreeding, the brains of this tribe of apes steadily grew more neurons through generations.

Traveling to far-away places, and seeing fresh surroundings, greatly stimulated the evolution of those brains. Australopithecuses grew to *enjoy* playing with fire. They thrilled to discover that a club-shaped tool is also a stunning weapon. They began to make meaningful grunts. A glimmer of *self-awareness* induced alertness and intensified emotions. Note that all this happened to the sense system of this particular ape that was an anomaly in apedom, while the other tribes remained chattering in trees in their home territories, or they descended and lurched around clumsily like gorillas.

Buzzing with ideas, and well fed in lush forests and grasslands, the peculiar ape's brain kept expanding with more and more neurons through the millennia until big watery cerebral hemispheres buried the small primitive brain deep in the skull. The "old brain," so called by doctors today, is the *hypothalamus*, the seat of sexual emotions and reproductive cycles, and wild animal outbursts of rage. This is the brain that we have from ancient mammals, and that came to them from reptiles and fish. In our "civilized" world the old ape brain triggers blind fury that overcomes the sober influence of the cerebral cortex in times of blind anger and panic.

The Australopithecus brain doubled its ordinary ape size around half a million years ago. Around 150 thousand years ago it had grown to "modern" size. Archeologists of our day, peering among the glaciers of the Ice Age, see the apes fade-out and glimpse an apeman with broad round shoulders and hairy arms holding a club and slouching beside a cave where he lives with his family. His brain is the size of an adult human brain—two and a half times bigger than the brain of the highest ape.

As a physical structure there is no essential difference between the dazzling computer in a man's head and the brain of a lobster, fish, or housefly. It is a matter of size and organization of nerve cells. As the aberrant ape got excited by tools and bonfires and high adventure, the cells of the cortex proliferated and proliferated until his brain was swollen with ten billion neurons with *a capacity for consciousness* that made him a human being.

Why did this happen only to the monkey line, and then to only one kind of ape? The last part of this question has already been answered: the mutation added extra neurons to the ape brain of Australopithecus, making him act in peculiar ways. After that the brain shaped up as the ape body developed. Long strong limbs developed like those of a trapeze artist. The hind pair of limbs invited this animal to stand erect and to stride and run with eyes sweeping the horizon. This animal was both nimble and able to spot food, cover, and enemies much quicker than an animal with belly close to the ground. The opposing thumb of the forelimbs, inherited from tree monkeys who grasp branches and fruits, turned out to be suitable for grasping tools and clubs —and later golf clubs.

Compare all this to what happened to the dolphin. Here is another mammal with the vigor of warmbloodedness. Its ancestors on land developed extra big brains for the animal—about 20 percent bigger than those of a chimpanzee. But dolphins had short limbs (now turned into flippers) on heavy bodies. This made them clumsy and gave them a tough time as land animals, so they returned to the ocean upon the continental shelves, where we find them leading quite intelligent lives compared to other ocean animals. They even talk to each other with whistles and squeals, and we can talk to them in the same way.

A Pacific bottlenose dolphin was caught, with four other young ones, in a school of some 80 wild dolphins gracefully gyrating among the waves in mid-Pacific. Segregated in a training lagoon, three of them were let alone to lead their lives, while one was trained by food rewards to turn and go at high speed to an underwater microphone, at the command of a police whistle. After ten weeks of training, the foursome were released far out in the open ocean. While they were still within earshot of the police whistle, its signal was sounded. The untrained three paid no attention. The trained one instantly hurried to the whistle to be fed.

Incidentally, training that dolphin's brain for memory did not change his old hypothalamic brain. When earlier attempts were made to lead him out to sea by towing a microphone which whistled meaninglessly at intervals behind a motor boat, the animal became greatly agitated. The moment he found himself in unaccustomed deep sea, his jaw chattered, his tail slapped the water, and he showed the whites of his eyes in terror.

Evidently, dolphins, which were originally land animals, returned to the sea after a good start toward becoming dolphinlike people. There they found a life of ease much better suited to fat,

94

heavy bodies and short limbs. When gravity was eliminated as a handicap, the dolphin could be lissome and gay, and become the world's most graceful diver and finest swimmer, able to maintain a speed of 25 miles per hour for an indefinite length of time. Dolphins had so few problems in the ancient cradle of life that they no longer had to stretch their minds, and their brains stopped growing when they were half the size of the human brain. It would seem that a medium-sized body with good mobility on land has the best chance of achieving consciousness.

When a cortex grows so big that it attains consciousness, it can make and break all sorts of new circuits in its electronic computer by the impulses of ideas. In this way, thinking, as opposed to living only by the ancient instinctive circuits, sharpens imagination, creativeness, and an awareness of the beauty of nature as revealed in sunsets and rainbows, flowers and green foliage. Out of all this comes a feeling of awe, love, and wistfulness—and a breadth of imagination that detects bonds between us and our pre-history ancestors who perfected our senses.

For this brain of ours still holds intact and in superb working order that old brain buried deep inside—the brain which lung fishes, aquatic earthworms, frogs, newts, and salamanders got from the plankton world and passed along to us. The human cortex, moreover, is no dry land invention. It is 95 percent water that is a solution of some of the most dynamic elements in the salty sea. It has hardly more substance than a jellyfish. In this sense, the human brain never emerged from the sea!

WHERE THE HUMAN BRAIN CAME FROM

So, organic evolution *in geologic time* bestowed a human brain on an ape. In the perspective of the evolution of the senses *in ocean time*, this assembly of ten billion nerve cells looks abnormal and isolated. Nevertheless, all the ancient planktonites —and all amphibians, insects, reptiles, birds, and mammals, whether with few or many neurons, or with brains scattered over their bodies as with earthworms or knotted in a head as with grasshoppers—used their senses in the same five ways to guide their footsteps through life by light, touch, odor, taste, and sound.

It is not precise to use the term "evolution" in speaking of the senses in the ocean. Sensitivity, as an inherent property of the living cell, did not evolve. Only receptors evolved to serve a particular kind of body and way of life. Receptors make some senses sharper than others in different creatures, but this process is not evolution.

No receptor of odor or touch is more sensitive than that of the waving hairs of one-celled flagellates. A fish has a keener sense of smell than we do. The feet of crabs feel more keenly than our fingertips.

Sense receptors have been redesigned to fit different forms of life, and their basic principles have been passed around among all ocean creatures.

What a contrast is the scene on land! Here, geologic changes resulting in swings of climates have repeatedly thrown plants and animals off balance. Species thrived for a few hundred years in a state of what ecologists call "dynamic equilibrium," only to be extinguished by coldness or dryness or lack of food—but not until they had bequeathed their immortal senses to new species with a different size, or toughness, and with new receptors and circuits for brain waves. Some of these became dominant until their turn came to be extinct.

Insects, which preceded mammals on the geologic timetable, have domed eyes by which they can see an arc of 180° without turning the head. With the three senses of touch, taste, and smell combined in their antennae, they achieved an acuity that makes them the dominant land life on earth with probably the best chance among all animals to inherit the land. Insects have such superior senses that they have the ability to dwell in a greater variety of environments than any other kind of land animal, and their size ranges from microscopic springtail to the New Zealand hoo-hoo flea, which measures 5 inches from tip to toe.

That is the story written in the Book of Stone, where pages are laid one on top of another with the upper ones labeled "later" and the lower ones labeled "earlier." On the other hand, in the original and biggest realm of life the plankton community still circles the globe, timelessly. The fantastic beings which oceanographers are staring at today are those which sharpened their senses with exquisite receptors for the animals of our world. Those in the ancient plankton which did not sharpen the senses for self-defense, and to obtain food and mates in the seething competition, vanished without leaving a trace. Yet the trail of the tingling five senses was never broken. Life did not start twice. The pervading senses of the eternal plankton beings link the human brain to the sensitiveness of the most ancient organisms on the planet!

9

The Ocean Endows Life with Sensitivity

The men were at the Columbia University Field Station in Bermuda with amplifying and recording equipment connected to a hydrophone in water nearly one mile deep. They were listening intently through headphones. After about two and a half hours of the vigil one held up a finger—"It should arrive about now"—and soon they heard it, a rumbling sound which developed and decayed according to an orderly pattern discovered several years before.

The Lamont-Doherty Geological Observatory scientists had exploded a 300-pound depth charge 4,000 feet below the surface of the ocean off the southwest tip of Australia. The sound waves, traveling four times faster than sound waves in air, crossed the Indian Ocean, passed the southern tip of Africa, headed northwest through the South Atlantic between Tristan de Cunha and St. Helena, and pulsed through a thousand miles of the North Atlantic, to make the rumblings in a microphone that had been planted over 4,000 feet under the sea near Bermuda.

What was it that physically traveled through the ocean to carry a rumble that fantastic distance? It was no material thing at all, but a train of sound waves—an excitement of H_2O molecules vibrating in a to-and-fro movement with the water fabric alternately compressing and expanding in a longitudinal direction, giving the forward thrust through the ocean.

Sound must have a material medium in which to travel and this medium must possess elasticity. Air molecules are delicately resilient, but in contrast to water they are heterogeneous, random, and loose, with the result that sound waves run through air only one-fourth as fast as through water—and they don't go far. A pistol shot can be heard only a mile or so. However, the capriciousness of air molecules makes them beautiful blenders of tones. Fish and dolphins could never enjoy a symphony concert.

Water substance is a lattice structure, a well-ordered system that transmits waves of vibrations of its molecules with a positive thrust. Sound waves are kicked along with the electronic energies of H_2O's. They hold a course with untiring vigor until they hit something firm, perhaps a headland or a whale, from which they echo and continue on at an angle. In an earlier chapter we considered in detail the wonderful fabric of H_2O's. What interests us now is the fine sensitivity of the ocean to sound pulses, and the way salt water transmits these with such strength and direction * that clicks, squeals, and whistles became vital communication for the survival and evolution of the colossi in the ocean and later the animals of our living world.

A subtle relationship between the fluid restlessness of the sea and the senses as an attribute of life bemused a class of teachers of child psychology at the Columbia University Summer School in 1956. They were asked whether the sun, clouds, lightning, wind, and sea are living. In their answers, 48 percent of this group of teachers stated that they believed one or more of these phenomena are alive. Other questions were more directly concerned with the senses. For example, "Ships are lost at the bottom of the sea. Do you think that the sea itself knows where they are?" And this question: "This pearl was once in an oyster in the sea. When the water moved over the oyster, did the pearl feel the motion?" In written replies one-third of the class said that they believed that the sea and the pearl were both sensitive.

Additional comments were: "Yes, the *chemicals* of the sea are aware." "Yes, the sea rules over lost ships and knows them to

---

* Salt solutions make water all the more sound sensitive. The saltier the water, the faster and clearer are the sounds.

be there." One of the teachers said the pearl felt "probably as a very young fetus might feel the water in the mother's womb." Over 40 percent of the class ascribed sensitivity to the ocean when asked: "Tides are caused by the gravity pull of the moon. Do you think the ocean can *feel* the pull of the moon?"

Those teachers of child psychology, confronted with subtle and searching questions, must have pondered the mobility and stroking nature of ocean currents, and then ventured to speculate that there may be a direct connection between the watery womb of life and the sensitivity of organisms. By their answers, which might have been ridiculed, they seem to have a prescience of what today's probing of the plankton world with electronic devices is discovering about the molecular basis of the senses and the enlivening effects of the chemicals in the salty ocean on organisms.

If such sentiments about mystical senses of the ocean seem unrealistic, what can be said about the sound of a depth bomb in the southern ocean picked up in the North Atlantic?

Note that both the band and the microphone were placed by the scientists about 4,000 feet deep. The ocean water is not all the same from top to bottom; it is stratified. In the upper 4,000 feet or so, called the thermocline, the temperature decreases rapidly to only slightly above 0°C. Below the thermocline the temperature decreases only slightly. The thermocline is of considerable importance in sound transmission. Below it temperature drops at a steady rate, reaching about 1°C at the bottom of the deep sea. Here are the haunts of weird monsters that carry luminescent lanterns in the perpetual blackness.* The zone just below the thermocline is the sanctuary for mating eels, and holds profound secrets of life. Space science techniques, directed down instead of up, are exploring this least explored area of our planet with "tritium oceanography"—radioactive tracers.

The upper layer that rests upon the tremendous foundation of cold, black water is only 3 percent of the mean depth of the oceans. Here, in comparatively warm water, in lush diatom pas-

---

* *The Galathea Deep Sea Expedition*, reported by the members of the expedition (London: Allen & Unwin, 1956).

tures, in the diurnal rhythm of day and night, the ancient plankton communities floated and drifted.

The sound waves are channeled on a horizontal course, and according to the laws of physics they follow the thermocline layer as they would follow a speaking tube. If you follow the course on your atlas you will be confused, because the 10,000-mile run between Perth and Bermuda appears to cut right across Africa. But if you stretch a string between these two points on a globe, you will see that the direct path passes south of Africa. That is, the direct path followed by the sound signal is a "great circle."

THE HUBBUB UNDER THE WAVES

We have just seen evidence of the fantastic sensitivity of ocean water to sound waves. They must have pervaded the plankton world from splashing whitecaps, bursting bubbles, jiggling beings, and long sound waves of roaring from earthquakes and landslides down the continental slopes. But there was profound silence in the plankton world, no sense of sound, until beings evolved receptors with a membrane of antenna that vibrates with the pitch.

Strong, fast-moving, far-ranging animals required more than one long-distance sense. They could not survive on sight alone. What was happening behind their backs? Then quite recently, on the threshold of geologic time, big, energetic animals found ways to put to good use the vibrant sensitivity of salt water. With long-distance hearing added to that of sight, the silent seething ocean produced the big fish, birds, and animals of our living world.

An early revelation that ocean depths are not "as noiseless as fear in a wilderness" came when the Woods Hole oceanographic ship *Atlantis* lowered a hydrophone a few thousand feet into the ocean near Bermuda in 1942 to listen to the sounds of distant underwater explosions. It also recorded a babble of squeals, shrieks, whistles, and ghostly mews and groans. Today's oceanographers are still trying to sort it all out, finding snapping shrimps, barking porpoises, and fast-talking sharks. Fish without lungs or vocal organs talk to each other by vibrating their swim bladders thus communicating with sound waves.

Fish sounds are on a different wavelength than anything our ears can hear. Theirs is indeed a foreign tongue. For a long time scientists thought fish were stone deaf. They have no eardrum, no organ like the marvelous "organ of Corti" in the human ear, which converts sound vibrations into nerve signals to the brain. Fish deafness was "scientifically proved" when a horrible clang, intolerable to the human ear, was produced underwater among fish. They paid not the slightest attention. But when a sound has *meaning*, fish react to it. They can be trained to scurry to food that is connected with a certain tinkle of a bell or a particular pitch of a whistle. Without eardrums or nerves for sound transmission, a fish has a sense of absolute pitch. The sound vibrations are received by fish scales all over the body and transmitted by bone conduction. Terrified at being hooked and yanked out of the water, fish protest vigorously in ultrasonic sound waves. Fishermen, who do not hear a squeak from fish that seem so mute when caught, can thus enjoy fishing the more.

When a shark is sighted off a bathing beach, the lifeguard's whistle summons bathers out of the water. Sharks are considered to have such an incredibly keen sense of taste that a few drops of blood from a cut finger puts a swimmer in dire peril. This is true for the close-in to a prey. The shark can taste a most delicate solution of molecules of blood (comparable to odor in the air) and zeros in on the prey by this sense. It was also very recently discovered that sharks possess long-distance hearing.

The native food for this fearful meat-eater is fish, not people. Through millennia they sensed their prey in the reaches of the ocean by using that wonderful sound conduction of water. The splash of fish at the surface draws them irresistibly. Splashing is the habit of many fish; a school of mackerel, swimming for dear life from a pursuer, jump out of the ocean by the hundreds, churning the water. Tarpon and tuna, which love to jump out of the water, are sirloin for sharks. When a man catches a bluefish, the struggling, splashing "blue" sends out ultrasonic sounds of panic that attract sharks miles away.

People swimming at a bathing beach when a shark is sighted are advised to use the breast stroke, not the crawl, to gain the beach with as little splashing as possible. The shark does not distinguish a human being from a tuna fish. Note that this animal with such a marvelous sense of hearing has no ears,* not even any bones. A shark is one of those ancient creatures with a gristle skeleton, before calcium made bones for modern fish.

The so-called "five senses" † are related to sense receptors which evolved with more modern animals to help them survive. Organs such as eardrums, retinas, taste buds, fingertips, and fossae (small cups about the size of a dime in the back of the nose) are similes of buzzers, light bulbs, magnets, transistors, etc., which use electricity in various ways.

Marvelous are the ways that sensitivity exerted its inherent power. In the lively melee of the ocean world, sense receptors became the chief features of evolving bodies, and signals between individuals—gestures, vocal organs, body colors, and markings —are the most conspicuous aspects of the living scene.

THE RESERVOIR OF SENSES TODAY

When we peer into the ocean we catch glimpses of events which took place in that four-fifths of life's sojourn on the planet during which sense organs evolved and bodies were shaped around them and field-tested by natural selection. In that two-billion-year operation, *sensitivity* played the leading role as the arbiter of creatures that have helped themselves to a portion of sensitivity and are vigorously trying out their personal sense

---

* It receives sound waves with its organ of equilibrium, called a "labyrinth."

† I use five senses for convenience, although more senses are recognized today. In addition to touch, taste, smell, hearing, and seeing, there is the temperature sense of cold and hot, the sense of pain, the muscular sense, the static sense of balance, an inner sense of time (the biological clock), and a sense of awe and wonder.

organs to deal with their environment. Those with senses in extra sharp focus that give them superior faculties for finding food and mates and for sensing danger are the extant ones. We know nothing about the astronomical numbers that failed to meet the requirements.

It is as though sensitivity is contained in a reservoir with a capacity equal to the totality of all living cells on earth. The physical store of this essence grew as organisms multiplied. In giant organisms it exerted great power in great muscles, but sensitivity never got sharper, never changed a bit. The psyche by which the blue-green algae insured their survival with touch and taste is exactly the same phenomenon as that which insures the survival of big organisms—although the latter acquired some elaborate organs for long-distance smelling, hearing, and seeing.

In the ocean community, where phyla are in a swirling tangle, organisms seem to have swapped sense organs. An outcome is that the sense organs of land animals all have their counterparts in sea creatures. The antennae of insects are the feelers of prawns, crabs, and lobsters. The entire sensory apparatus of mammals, including man, came from fish. Centipedes and some earthworms have peculiar light-sensitive spots called ocelli (proto-eyes); so do shellfish. The human eye and the unblinking, hypnotic eye of the octopus are structurally alike in almost every detail.

The brain of the cuttlefish is a fist of nerve ganglia enclosed in a skull made of tough cartilage. The fist is divided into lobes, each of which governs a particular set of emotions and reactions. One lobe controls the emotion of alarm and releases a jet of black ink into the water, behind which the cuttlefish makes good its escape from an enemy. Another lobe controls cuttlefish memory. Another, the biggest lobe of all, controls sight. This description could also apply to the lobed human brain. Of course we use different reactions to the same senses. Instead of emitting a smoke screen when alarmed, we tense our muscles and prepare to run or fight.

The cochlea of the human ear is a helical spiral like that of a snail shell. Evolutionists say that we inherited our ears from the gills of fish—when eardrums were installed in the spiraling bones, breathing organs became hearing organs.

The static organ in the ear gives us the sense of being right side up. This exquisite sense receptor is a hollow bony wall, lined with soft hairs, each of which is a highly sensitive nerve cell. A limestone pebble rolls about, like a ball in a bowl, striking against the hairs signaling the position of the body with respect to gravity, telling the animal as it tilts or is upset how to restore an upright position.

Even more surprising, this identical device was used by jellyfish epochs before there were any bony fish. At eight points around the edge of the jellyfish umbrella there are the little hollow balls lined with sensitive hairs struck by the rolling pebbles. This is surely a critical situation in which to perfect an organ of equilibrium. It is hard to imagine any animal more tossed around and upset by rollers and whitecaps, than jellyfish.

I repeat that sensitivity has never changed in the least since the first living cell appeared on earth. All the wonderful apparatus of the "modern" animals—including our own ears, noses, fingertips, and taste buds—have not made any sense sharper, have not created any new kind of sense. What has evolved are sense organs and more complex systems of response.

What about ESP? That would be better called ISP (Inner Sensory Perception). An instinct that produces a vague echo of consciousness in the human brain seems to me in no way superior to the sense of a one-celled ameba which, without any sense organ and not a single neuron, makes its way continually detouring around obstacles and creeping toward molecules of food, and sensing the slightest variation in the water solution.

The late Professor Sinnott of Yale University, in his book *Cell and Psyche*,* points out that this behavior of amebas is quite comparable to that of the higher organisms equipped with the latest sense perceptors and motor nerves. He adds that even plants, the most stationary of living things, respond to the touch of light and moisture by growth movements or turning

---

* Edmund W. Sinnott, *Cell and Psyche* (Chapel Hill: University of North Carolina Press, 1950).

toward the direction of the light rays. Moreover, with no nervous system, the stimulus received in one region of a plant can be transferred to a distant spot. For example, a slight touch at the leaf tip of the "sensitive plant" (*Mimosa*)—or merely blowing on it—causes the leaflets progressively to fold their two sides tightly together until the whole leaf bends down sharply.

We noted in the last chapter that pigments in blue-green algae are as sensitive to wavelengths of light as pigments in the retina of the human eye. Wavelengths of visible colors are far, far smaller than anything we can consciously see. The smallest visible dimension is about 1/25 inch (one millimeter) and visible light waves are around 1/60,000 of that. The chemical sense of the earliest known organisms in the world was keen enough to detect and put to use light waves with such dimensions!

When in the course of eras the motes of blue-green algae and bacteria evolved the magnificent protozoa (life was still only single-celled, but larger and reaching out with hair projections, and able to row around with bristles, like banks of oars), the little old ancestors handed along their sensitiveness. The elegant Paramecium, a giant among the single-celled animals, corkscrews through the water, revolving on its longitudinal axis, with its rows of bristles bending with the rhythm of rows of wheat waving in the wind. Note how those first sense organs were shaped and controlled by the feel of water.

SENSES ARE SHARPENED FOR INDIVIDUAL USE

We return to the melee under the ocean waves where bizarre creatures are honing their senses. Each kind of creature sharpens its portion of sensitivity from the reservoir in its own way to locate and attract its kind of mate, to guide it to its favorite food, and to detect the approach of its personal enemies.

Emphasis on one way of using sensitivity gave some ocean creatures, far below us in the Tree of Life, much keener senses than we have. Moreover, this spearheaded organic evolution with such force that it finally drove the Arthropods and Chordates out of the water, on to the land. Thereafter, focusing on a particular sense for a particular need sharpened particular senses among land animals.

A dog has a much keener sense of smell than its master. A hawk floating a thousand feet in the air can spot a rabbit half hidden in the grass. Did you ever watch a robin after it has prowled on your lawn and found a worm hole? The bird stoops with its head close to the ground—it is listening for the stir of an earthworm underground that will tell the robin whether the worm is about to emerge.

Musca domestica, the common housefly, of the same phylum as lobsters and crabs, has a pinhead brain—yet so keen are its senses that its skill as a flyer is fantastic. Try to smite the pesky thing. The chances are that it will sense your descending hand and act with such dispatch that the next time you catch sight of it the fly will be buzzing around your head taunting you. You might as well enjoy the greatest acrobatic act on earth. The fly makes turns in the air too fast for the eye to see, it darts quickly up and down, describes figure-eights, takes off from a standing position, zooms down to land at full speed without even a stagger to adjust to the impact.

People can sharpen particular senses with practice, even substituting one in place of another. Here is a quotation from Helen Keller: "The sense of smell has told me of the approach of a coming storm hours before it arrived. I noticed first a throb of expectancy in my skin, a slight quiver, a concentration in my nostrils. As the storm draws near, my nostrils dilate, the better to receive the flood of earth odors which seem to multiply, until I feel the splash of rain against my cheek. As the tempest recedes farther and farther, the odors fade, and die away."

Honeybees exhibit senses sharpened for a wondrous way of life. When flowers open in the springtime they beckon the bees with odors and colors. Bees, even three or four miles away, use keen senses to hit the pinpoint targets in the vast landscape, first, long-distance *sight*—bees have domed eyes which can see in an arc of 180° without turning the head. The plants signal with colors and vivid patterns, or perhaps a swaying tassel, or sparkling nectar. *Touch* guides the feet of the bee after it alights, sweet *taste* identifies the nectar. When it has a full load of pollen or nectar it heads for home, using the 12,600 lenses of its domed eyes to navigate by the angle of the sun rays.*

This shows an interplay of sense signals between the Plant Kingdom and the Animal Kingdom!

SMELL—THE SENSE OF SEX

In 1869 a scientist imported a few mating couples of gypsy moths to Medford, Massachusetts, in order to crossbreed them with silkworms. One pair crawled across the laboratory windowsill and escaped. A few years later the New England states had to be quarantined against gypsy moths which were killing thousands of acres of shade and forest trees by eating their leaves. How could they breed so fast and so far?

Scientists contrived all sorts of experiments to find the answer. One was a carefully planned smell test on Cape Cod where there were no gypsy moths at that time. A female moth "in heat" was put in a cage suspended from a branch of pitch pine. A male moth, marked for identity, was released two miles away. After allowing for the rate of a moth's fluttering course, observers beside the cage were delighted to see the male arrive on schedule and ardently knock at the bars of the cage.

Jean Henry Fabre, the Frenchman who wrote a ten-volume work on insect life, and whose laboratory is now a national museum in France, tells this story about his early collecting days. He was thrilled one day to come upon a cocoon of the seldom seen emperor moth. He watched in awe as the creature unfolded its gorgeous wings. He breathlessly made notes as it

---

*For a fascinating account see *Bees: Their Vision, Chemical Senses, and Language*, by Karl von Frisch (Ithaca, N.Y.: Cornell University Press, 1950).

flapped around the room, in which the windows were carefully closed. Then he carefully deposited his treasure in a bureau drawer for the night. Soon after retiring that first night he was aroused by sounds of scratching. Males had arrived and were knocking on the window glass!

Fabre continued his research with windows closed and the empress in her bureau drawer each night. During the next few nights, scores of those very rare emperors, summoned from near and far by signals from the bureau drawer, crowded in a frenzy at the window of the room. In a frenzy, too, were Fabre's family and the whole neighborhood, while he pondered "a perfect nonplus and baffle to human understanding"—how did she summon the boys?

The renowned naturalist probably knew that female moths call mates by the sense of smell, but 80 years ago he could not know that odor is transmitted in a submicroscopic bit of matter which can pass through keyholes and cracks of a loose window sash and go long distances with undiminished potency. That knowledge had to wait for the molecular biologists of today. Even as this is written, scientists are astounded by the miracles of odor molecules, particularly in the water.

The subtleties of senses used in finding a mate began with the localized sense of touch, and were extended a distance by sight, taste, hearing. Finally, odor, the dominant one of all the senses, was perfected for long-distance communication. It could call across any distance near or far, even through miles under water, to guide sexual reproducers. It may have been an evolution of the original chemical reactions of taste.

The odor particle now bears the elite name, pheromone (Greek for "molecule that carries odor"), but its mystery is still intact. It evolved in the primordial ocean to meet the requirements of the more elaborate animals with wider and wider distribution. After life emerged from the ocean, pheromones still required a watery environment to operate. This exceedingly minute particle can be carried in droplets of invisible water vapor in the air. To produce odor, a substance must be moist. And this unique pheromone item must be dissolved in a fluid spiced with enzymes to yield its signals. For us, this is a function of the mucous in our noses.

Baron Wolfgang von Buddenbrock, world-renowned authority on animal senses, muses: "Was sex promoted by smell, or was smell promoted by sex?"

10

The Two-Animal Animals

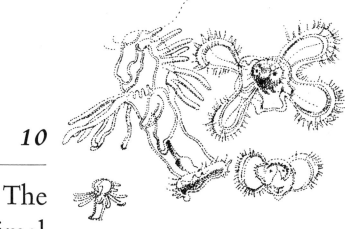

It is hard for us to imagine how awesome was the mystery of the Ocean Sea when Columbus sailed through the Pillars of Hercules and disappeared into the sunset. Most people still believed that the world was flat and their thoughts were colored by the belief that the ocean was an uncrossable stream of water flowing around the lands and blending with the sky beyond the horizons. Dante, who was widely read in those days, related in his *Inferno* how Ulysses sailed into the vast void beyond the horizon to die. There was no idea that the earth is a globe; the general opinion was that if any lands lay beyond, they must be unnatural lands inhabited by unearthly beasts and giants.

Today, with sonar, radioscopes, and water-tight cameras, explorers are crossing equally mysterious horizons under the keels of ships, in a shadowy world teeming with beings more imaginary, more "unreal" than the giants and beasts of ancient lore. I refer particularly to the two-animal animals * called larvae and adult, one animal transforming into a second utterly different in its shape, size, and way of life.

---

* I use the term "two-animal" for convenience, although in a few cases an animal assumes more than two forms. Indeed, as many as eight forms have been detected, counting all stages of metamorphosis.

Of course, examples of this weird larva * phenomenon are common in the everyday world around us, such as the caterpillar-butterfly, the grubworm-housefly, the tadpole-frog, which are taken for granted as a natural way of life—like babies being born, and buds turning into flowers, and flowers into fruits. What makes the two-animal animals in the ocean so exciting is the light they throw on the baffling enigma of life on this planet and on a crisis of gravity which occurred in the dim depths of the primeval seas.

In preceding chapters we noted how the diatom "green leaf" achieved an efficiency for photosynthesis that has never been surpassed. It nourished elegant protozoa with rows of *cilia* (vibrating hairs for swimming) and *flagella* (long waving hairs) that collected bacteria and other specks of delicious food, and microscopic protozoans even had keenly sensitive organs such as eye-spots and gullets. How could living things be better designed to become the permanent population on the globe?

The innate urge of living cells to eat and reproduce gave the protozoa insatiable appetites that resulted in explosive multiplying. At the same time blooming diatom pastures installed ever more sunlight energy in the sea water, now also enriched by nourishing broth dissolved from the protozoans' dead bodies. Then one-cellers possessed with excess vigor seized each other and clung together, making all sorts of aggregations which united their energies.

As long as they were buoyant and could swirl and travel in currents, the clumps of one-cellers could prosper as well, perhaps better, than individual protozoans. In the luxurious Era of the Protozoa, the chief requirement for survival was buoyancy for rising and sinking with the vertical day and night rhythm of the photosynthesizing diatoms, for plants were then, as they are today, the food suppliers of life. Worldwide travel in ocean currents gave the buoyant beings distribution that relieved overcrowding. The incalculable benefits derived from good distribution with ample room to operate is depicted in the Plant

---

* The Latin word *larva* means a mask, specter, scarecrow. Early observers viewed tadpoles and worms as specters of frogs and moths.

Kingdom, where things firmly anchored to a spot achieve wide distribution and prosperity with windblown pollen and seeds.

Certain peculiar beings to be seen today show us in detail how protozoa, in the buoyant jubilee of that era when they were the "important" beings on earth, may have begun to shape up many-celled bodies. For example, there is the Pandorina—Little Pandora—who, for the last billion years, has been taking the first step toward visible metazoan life. The name has the lore and poetry of so many Latin names of plants and animals.

In Greek mythology Pandora was the first mortal woman, and on her all the gods bestowed great gifts. Venus gave her beauty; Mercury gave her persuasion; Apollo, musical talent, and so on. Then Jupiter sent this ravishing woman down from heaven as a subtle punishment for a reckless man named Prometheus, who had stolen fire from heaven. Jupiter intended that Pandora should make a lot of trouble for Prometheus and all mortals. She did. Unable to curb her curiosity she uncovered a tempting box and set free all the plagues of hapless mankind.

Our Pandorina's body is composed of only eight cells embedded in a blob of jelly. This is not an organism, it is a team of one-cellers working together, which gives it an advantage, with a little more energy, a little longer life, than the unattached ones. Similar teams of one-cellers with any chance number of individuals gave the Era of Protozoa a "new look," lifted the life power of individuals to a higher level. They jigged and swirled and circled the globe in ocean currents like all the other protozoa. They were merely a bunch of one-cellers sticking together while enjoying all the emoluments of the age. With no menace of food shortages, and no fear of becoming the prey of giants, we might consider that "The Kingdom of the Protists," as zoologists call it, was the peak of carefree life on earth.

Biology students staring at protozoa through microscopes often see a grandiose globe, lumpy like a golf ball, float into view. This is the famous Volvox, made by a thousand flagellate protozoa united in a ball with a gelatine coating. A mysterious sense of togetherness possesses these protozoa, as seen in the way they form a perfect sphere and wave their hairs in unison to make their ball slowly revolve as it drifts around.

Volvox can live on dead protozoa as an animal does, and also, with patches of green chlorophyll, it can use sunlight energy to make its own food like a plant. Even more sensational, this plant-animal reproduces itself in two ways. Some of the protozoa that compose the ball multiply by the ancient, simple method of dividing in two, and others by a new-fangled ritual called sexual reproduction—much too complicated to have a future!

In the first instance, single-cellers break away and move into the interior of the hollow sphere, where they form hollow groups called "daughter colonies." When the so-called daughters grow up they swarm upon their mother and tear her to pieces—a ceremony of great importance, as we shall see in a moment.

The other way that a Volvox reproduces occurs when certain individuals composing the Volvox sphere acquire male and female characteristics. The "sperm" break out and whirl off, each vigorously waving its two flagellate hairs as it goes in search of a mate. When cells of opposite sex collide, they cling together and "breed" a baby Volvox with two cells for a body. A chain reaction of dividing cells quickly makes a beautiful new Volvox globe.

An unprecedented problem looms for a Volvox with thousands of cells in its body. Indeed, this is the crisis which threatened the survival on earth of the many-celled animals. Volvox has lost much buoyancy; it is ponderous. It can keep going for a while with the members of its ball furiously waving their hairs —but it cannot swirl lightly and swiftly with the one-celled multitudes. It can survive in that time-honored dance of life only by fragmenting. For this, the frantic violence exhibited by the daughters tearing their mother apart is a great help. As for the daughters, and the sex cells, they are lissome and light-weight, and the latter can carry on their exiguous sex life with buoyant agility.

Up to this point all life on earth was as independent of gravity as drops of water. The next step in evolution—the creation of true metazoa—will lead to structures of weighty shells, bones, organs, heads, tails, fins, and limbs. Metazoa cannot multiply apace in multitudes, as do microscopic beings. They must have time to grow up, develop sex organs, become adult to reproduce. Meanwhile, struggling mightily to hold buoyancy, they slowly sink to the bottom, away from food supplies of buoyant diatoms and their single-celled ancestors in the upper layers of sunlit water.

On the face of it, inexorable gravity presented a problem seemingly impossible to meet. Life on earth was condemned to be microscopic forever—ruled over by creatures like Pandorina and Volvox. But wonderfully, the versatility of water was installed in its creature, the living cell. The high hurdle which stood at the top of the Era of the Protozoa was surmounted dramatically by the invention of two-animal animals.

THE INVENTION THAT MADE LIFE VISIBLE

Theirs was a well-kept secret—how larvae, once upon a time in the ancient ocean, gave rise to the colorful, flying, swimming, running, mating, hungry, fighting creatures of our world—which outwitted the law of gravity.

Through history people have wondered at and delighted in oddities of the sea such as barnacles, crabs, starfish, jellyfish, sea anemones, lobsters, clams, and snails at the edge of the sea—and iridescent fish that leap from the waves, and whales and porpoises which look like black wheels turning in the ocean.

In more recent times mariners and fishermen regarded with amazement the peculiar tiny animals which bloomed with the seasons and colored the water. They were always rushing around, possessed with unrestrained energy.

These were fairytale creatures. Fish and shellfish supplied delicious food; sharks, octopuses, and whales could be dangerous enemies; but for the most part, the hordes of tiny ones had no more relation to life on earth than beings on another planet.

At the turn of the century the science of oceanography began to shape up as something more than mariner charts and fisheries. Some biologists became immensely curious and excited over the "funny" little creatures which silently and perpetually drifted around the world in ocean currents. They were certainly no "inhabitants of another planet" out there in the ocean, they must hold some profound secrets of life on earth.

How did they reproduce? They could not multiply by cell

division, by simple splitting in two—they were metazoa with eyes, legs, and organs, although some were hardly bigger than a grain of sand. Moreover, they were not observed to be male and female breeding their kind by sexual reproduction.

Laboratories were established on seacoasts, and gradually the realization grew that these hordes of live beings whose spectral clouds dilate and stream in the ocean are *larvae*. Oceanographers put them in tanks of seawater, through which salty, organic solutions from the ocean circulated, and watched them as they swam around in a desperate competition for food. They grew bigger and heavier, until their frantically beating legs could not match the force of gravity and they fluttered downward. While in repose on the bottom, miraculous transformations took place. Their organs and appendages steadily turned into other kinds of structures, the shapes of their bodies became altogether different —they turned into jellyfish, oysters, barnacles, shrimp, crabs, and other familiar ocean citizens.

These first discoveries of these two-animal animals came in a decade when adventure in unexplored areas was in full swing. Theodore Roosevelt had plunged into "darkest Africa"; Robert E. Peary attained the North Pole; W. H. Hudson was in the southern pampa where wild horses and rheas frollicked around him and astonishing birds soared above. The two-animal animals invisible under the waves were too tiny to compete on that level of public excitement.

One champion of ocean larvae turned up. The prince of Monaco was an adventurer with a lively imagination and lots of money. When in 1911 France assumed the political matters and the defense of his half-square-mile principality, the Prince built a plush gambling casino. Soon, booming revenues enabled him to equip a fine yacht with tanks, trawling nets, and devices for oceanographic exploration, and also to build a superb marine institute on a rocky bluff overlooking the sea. This was the ideal location for pumping water from under the blue waves of the Mediterranean through tanks, and keeping them well supplied with life-giving solutions of organic matter and an endless parade of little animal forms.

The Marine Institute still stands beside the famous Monte Carlo

Casino where today Prince Ranier and his Princess, Grace Kelly, can enjoy adventure on the frontier of the most mysterious realm on earth.

Almost all the denizens of the sea—jellyfish, octopus, oyster, clam, snail, starfish, sea urchin, sea anemone—are two-animal animals. Curious creatures are continually being added to the list and given picturesque names that suggest their sporty forms and colors—Venus's-girdle, moss animal, lamp shell, sand dollar, comb jelly, serpent star, brittle star, sea lily, sea cucumber (the last two are animals despite their names).

The most ancient, and perhaps the original two-animal animals are the worms. That multitudes of species of worms are charter members of this legerdemain society is hardly appreciated because they hide in sediments and rock crannies. They seem to be immune to crushing water pressure. An acorn worm has been photographed dredging a spiraling path at the bottom of the Marianas Trench, the deepest spot in the ocean, seven miles below the waves of the blue Pacific, near Guam. Worms also burrow into the rocks of low-tide pools.

All the above are the *adult* forms of their two-animal selves. They are big and easy to see, compared to their minute transparent larvae. Their hunger has abated, they have developed elaborate sense organs, and their paramount assignment is to find mates and reproduce their kind with particular rituals.

Since the adults are conspicuous and slower moving than agile larvae, these animals become easy prey in the course of evolution. The very reason for their existence, the survival of the species, could be defeated. But natural selection gave them ingenious devices for defense. Their chief instinct is self-defense —to be let alone just long enough to produce the next generation.*

This inherent drive of live stuff to survive and reproduce has neither abated nor altered from the first single-celled inhabitants of the planet to the many-celled animals, to all territory defenders, to tribes, to societies, to countries. *The overriding*

---

* In some instances, defense weapons were used to hunt prey. For example, the stinging cells of the monster Portuguese man-of-war, and the shocking discharge of the electric eel.

urge to grow up as fast as possible and reproduce as fast as possible created the amazing two-animal animals.

At the instant a larva is born from an egg its legs snap out and start to whirr, and it goes darting around without a moment's delay. It has extra large eyes and fully formed, sensitive antennae. The legs had been wrapped around its body coiled inside the egg. The larvae are little, light, and buoyant, thwarting the down drag of gravity. This automatically gives larvae the same advantages as protozoans for roaming in diatom pastures and enjoying ever fresh vital solutions of salt water through worldwide distribution in ocean currents.

However, the larvae are not just traveling companions. As swimmers they are far superior to single-cellers waving their feeble hairs. The larvae are the masters of the ocean world, while the multitudes of protozoans and diatoms are their feast. The newborn larvae eat ravenously to grow up as fast as possible. In the shortest possible time they must develop body equipment, and above all reproduction organs, to hand over to their adult selves, who will put them to use.

PLETHORA FOR SURVIVAL

Sir Alister Hardy, professor of zoology at Oxford University, emphasizes the distribution advantages of larvae animals in the ancient ocean with these words: "The larval stages are as much adapted for dispersal as are the seeds or fruits of plants." *

An excess of numbers is a vital principle of life. Every living thing on land and sea today uses this principle at some time in its life cycle. A single oyster, an inert blob in a rugged shell, releases some 500 million eggs into the surrounding water. The number of human sperm in a single ejaculation averages 240 million. Pine trees and ragweed erupt with geysers of pollen. Countless seeds rain from trees and bushes, and gush from grasses. Where myriads perish, a tiny residue survives to per-

---

* Alister C. Hardy, *The Open Sea* (London: Collins, 1956). This book, a work of art with its fine illustrations, has been a chief reliance in my account of plankton and larvae.

petuate its kind. The mighty life insurance of plethora, and easy transport over large areas of potential life-space, was as important as buoyancy to the larval animals. Small enough to plug into the dynamics of the protozoan world, the larvae created organ by organ, detail by detail, the animals of our living world.

At first the scientists, even as you and I, were fascinated by the revelation of the ocean larvae. The electron microscope and other techniques are still discovering even more strange forms of life hidden in the ocean. At first studies were aimed at classifying the larvae creatures as though they were a class of beings distinct from the everyday world. As it turns out, they are the very trunk of the Tree of Life.

As the studies continued and the peculiar animals were paired with familiar jellyfish, starfish, crabs, snails, barnacles, oysters, and so on, the hordes of larvae which bloom and billow beneath ocean waves became recognized as a living exhibit of what went on in the ocean in an era long before geologic time, when the species we see on earth today were being created. It was as though the curtain had risen on a "new" act in the drama of the evolution of species, revising some long-held beliefs about how such utterly different kinds of organisms—fish and birds, fleas and elephants, mice and men—turned up on this planet.

The classical assumption based on Charles Darwin's great mechanism of evolution, "survival by natural selection," is that it operates primarily in the gene codes of mating adults—genes spell out the characteristics of a species and hold it true to form in a line of descendants. However, many little accidents happen to genes. Some may be knocked out by ultraviolet radiation, or mishaps may occur in the pairing of male and female genes in mating. Through epochs of time the favorable among these changes (mutations) so alter the characteristics of a breeding population that a new species develops.

Those characteristics which peril survival eliminate their animals, those which are beneficial give their breeders advantages for survival, and the line of descent proceeds—from backboned fish to reptiles, which branch off to birds in one direction and mammals in another.

The study of embryos underscored this theory and led to

what seems to be a brilliant insight—that the development of an individual, from fertile egg to adult, recapitulates the evolution of its phylum. It was a fascinating discovery that the human fetus has gills reminiscent of its fish ancestry, and the backbone has a tail that hints its reptile past, not to mention the vigorously swimming tadpole-like sperm. This wonderful concept bears a solemn scientific label—*epigenesis*. It became almost "gospel" of biology.

The fossil record seems to confirm this theory that evolution of the whole pageant of life on earth arose from breeding adult populations, which in the course of geologic time—when they were confronted with the perils of changing climates, ice ages, mountains rising, and lands fragmenting into drastically different habitats like deserts and swamps, tundra and tropics—the mutations which helped survival made dramatic evolutionary changes. Thus the world was left as we found it, with animals fitting their environments, the kangaroos on the Australian plains, ostriches in sandy Africa, polar bear among Arctic icepans, buffalo on American prairies, alligators in the Everglades, and the anthropoid apes up in trees.

How, we may ask, do the hosts of larvae in the ocean affect that kind of evolution?

THE GREAT SECRET OF THE OCEAN LARVAE

The fossil record is the solid evidence of classical organic evolution. It compares teeth, skulls, and skeletons of adult animals and, on the timetable of the rock strata of the earth's crust, it discloses how animal bodies evolved.

But note that the beginning of the traditional fossil story more or less coincides with the emergence of life from the ocean. The fossil story gives the impression of starting quite suddenly, with a good many creatures identifiable with those of our time already elaborated, seemingly well equipped with what it takes to live in our world. The geologic calendar says that this kick-off of organic evolution was around the beginning of the Cambrian period, 600 million years ago. What happened before that? What about the evolution of life in the timeless ocean?

Answers to that question awaited the new perspectives and dazzling arithmetic of the space age, which can probe both the outer universe and the cosmos hidden under the ocean waves. The tale of larvae, which came into existence on the primordial globe when one-celled organisms were the only kind of life on earth, makes the story of fossils in rocks only a postscript to the evolution of life. The latter indicates the relationships of different kinds of creatures *within* their respective phyla. The ocean larvae whisper to us about the kinship of worms and oysters, of octopuses and people.

The pioneer detective of the larvae's story was Walter Garstang, Professor of Zoology at Oxford University, in 1920. He was an extraordinary combination of scientific skills, artistic excitement, philosophic insight, infinite patience for delving into the inner sanctums of ocean larvae which were his passion, and a writer of comic verse. His students never forgot his lectures, but his findings that larvae were the architects and founders of organic evolution on our planet were never put in a book, and they became all but lost until recently.*

I am indebted to Alister C. Hardy for this sample of the way that Garstang reported to his students:

> *They feed and feel an urgent need to grow*
> *more like their mothers.*
> *So sprout some segments on behind, first*
> *one and then the others.*
> *And since more weight demands more power,*
> *each segment has to bring*
> *Its contribution in an extra locomotive ring.*
> *Then setose bundles sprout and grow, and the*
> *sequel can't be hid:*
> *The larva fails to pull its weight and sinks*
> *—an Annelid.*

---

* An interesting parallel was the work of the monk Gregor Mendel, who crossbred sweet peas in his garden and discovered the mechanism of heredity in 1866. His findings were overlooked for 35 years.

Garstang had described a buoyant larva reaching the point when it is ripe for reproducing and then turning into its adult, a comparatively heavy worm. Here is another of the professor's scientific reports about larvae:

The simple stage with prototroch is by-passed
in the eggs,
And each when hatched had three good pairs
of parapodial legs;
With these it paddles on in jerks, and
could it change its skin
Would almost be a Nauplius, *its very near*
of kin.

Nauplius is Greek for "little ship under sail." Nauplius does not sail on the surface, however, but under water, rushing with great gusto in all directions.

The Nauplius is just barely visible, about 1/16 inch long. It has no head, six bumps on its underside, two hornlike projections at the two front corners of the body, and one eye in the middle between the horns. As it feeds and grows, it continually breaks out of its shell and forms a little bigger shell to fit. It expels its excrement with each castaway shell, an arrangement that makes it unnecessary for Nauplius to have organs for that purpose.

There are minor differences between Nauplii, but on the whole they are all the same kind of organism, and they all conspicuously resemble each other. At first sight, Nauplius is an uncomplicated, insignificant item with its innards simplified, and its outward appendages reduced to a minimum. It would hardly catch the eye of collectors of plankton life. Yet it has a surprise up its sleeve—the "sleeve" being the inside of this tiny body, where the most important "vital organ" in every living thing on earth, the gene code, is ready to produce a startling diversity of the most bizarre species and the biggest numbers of ocean animals. The Nauplius is the larva of the *Crustacea.*

As the time approaches for this thing to turn into an adult, the bumps on its underside elongate into strong, stumpy legs with which it swims strenuously. The horns get long and slender and sensitive, to be antennae; the single eye divides, and its halves slide apart to give it a pair of eyes. Then, after its last molt, Nauplius becomes a copepod (the most numerous of all animals on earth)—or it may be a barnacle, water flea, prawn, shrimp, crab, crayfish, or lobster.

Judging by results, the Nauplius style of body is the most successful design ever invented for the first stage of two-animal animals.

The multitudinous progeny of Nauplius evolved branches of the Crustacean phylum which pervade our world today with spiders, wood lice, centipedes, millipedes, sow bugs, ticks, scorpions, cockroaches, beetles, grasshoppers, dragonflies, ants, wasps, bees, indeed all insects—of which 75 percent still lay their eggs in water. That is the accomplishment of a Nauplius larva!

In his entertaining way—and perhaps without fully realizing its import—Professor Garstang announced a theory about larvae that gives standard organic evolution a huge new dimension. This does not discount Darwin's "survival by natural selection"; indeed, it makes Darwin's perception of the origin of species even more profound by removing the chief theater of the drama of evolution from rock fossils to the immense and far more ancient ocean.

Consider the circumstances: the immensity of the ocean, the global currents, and the incessant mixing of vital salt solutions. Then, as hosts of protozoans and larvae proliferated and died, the seas became a highly nourishing olla prodrida. And so it was for a billion years, perhaps two billion years, while the larvae danced and swirled and themselves evolved by little accidents to their gene codes.

The egg and sperm cells of the big animals which left fossil imprints in the rocks were so deeply imbedded and well protected against many hazards, that vital mutations could not happen every instant. The ocean larvae are far more vulnerable. Their tiny, transparent bodies are soft and pliant. They jig through changing pressures and light intensities, and they present themselves at all angles to their surroundings. Their generations are astronomical. Their time was akin to infinity—anything could happen to their genes, even "miracles."

The bodies of those ocean larvae could not be drastically altered. Their function was strictly circumscribed—to eat, grow, and start an embryo. They could not get much bigger without losing buoyancy. But what could and did happen to them was that their *gene codes changed*, and this alteration took place without corresponding change in the size and nature of a larva. Yet this was a key change, which was destined to create all sorts of innovations in future ages, when larvae passed their altered inheritance to their adults along with the rudimentary reproducing organs.

This fundamental evolution could leave no fossil record. It operates on a molecular scale, and it is hidden deep in the ocean. Moreover, the ocean larvae dissolve when they die. Those which have shells made of limestone or silicon deposited carpets on the ocean floor. But ooze carpets are not rock strata which can bear the imprints of ancient animals.

Oceanographers today are finding tantalizing clues to the climates of past ages and are surmising about ancient life in the sea by studying the depths of the oozes on the ocean floor and their molecular compositions. In this connection, dramatic geologic events such as drifting continents are being studied. And there is strong evidence that heat currents, called *convection currents*, are revolving in a colossal cycle in the outer crust of the earth, and that this carries down the deep ooze carpets on the abyssal floor—they are being swallowed by the crust.

Thus, the ocean oozes telling their secrets of bygone ages are identified with the Quarternary and the Tertiary periods, the most recent on the geologic calendar. The ocean world of the larvae which we have been considering was hundreds of millions of years before that. How awe-inspiring—that two-animal animals, like those which created metazoa in the primeval ocean, are still alive and in full operation!

Cross section of a giant conch reveals the rhythm of a breaking wave.

Strawberry leaf exudes water drops in process called *guttation*. This water squeezed out of teeth of leaves is often mistaken for dew.

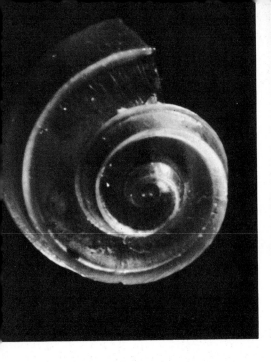

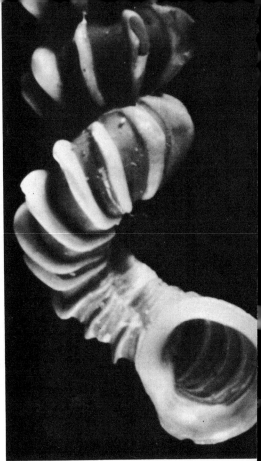

The sculptures of shellfish bear the ripples, spirals, and curves of the water in which they are formed.

Just after sunrise, when a spider web is spangled by dewdrops, you can see how the spider spun its web with the spirals of a vortex.

The Ice Age leaves its imprint where an Arctic river carves the continental rock. (*Rex Gary Schmidt, Courtesy Bureau Sport Fisheries and Wildlife*)

A snow blanket keeps the ground moist, protects seeds, roots, and soil creatures from freezing temperatures.

In early spring the warm touch of sunlight sets the streams running, awakens life of woods and fields quite suddenly.

An awe-inspiring exhibit of water on our planet. This hoary iceberg is made of snowflakes from blizzards of the Ice Age. The caverns show the rush of water where heat from the pressure of the ice cap melted the base ice, created mighty rivers.

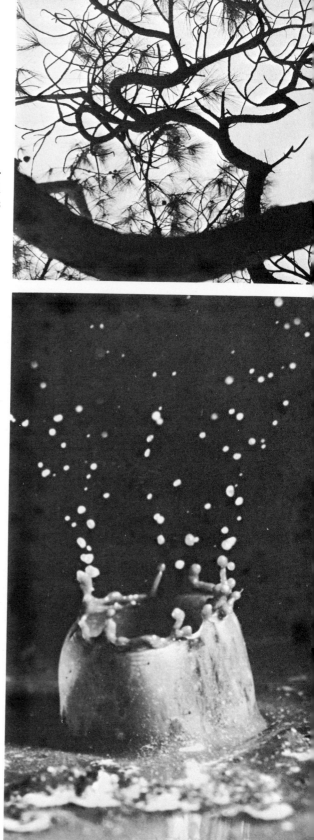

Tree branches are like hoses—water flows through all their twists and turns. (Picture shows branch patterns of Torrey pine.)

Impact of raindrop on the ground turns it into a tiny volcano that erupts with microscopic drops.

The reflective talents of water have given life to many a landscape. (*Carl A. Koenig*)

11

Tape Recordings About Sex in the Ocean

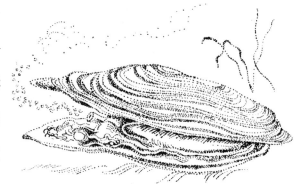

Headlines about S-E-X proclaim this subject as though it were the big news of the mid-twentieth century, and as though its dazzling possibilities had just been discovered. Of course, in this matter, we are different from all other members of the Animal Kingdom only in that we think about, write about, and exploit sex as an *elixir vitae*, while they do not. Otherwise they are just as possessed by sex and its results as we are—and, indeed, instinctively and impulsively, other members of the Animal Kingdom focus on sex just as fiercely as people, and many dress in finery and specify elaborate ceremonies in wooing.*

I do not have to expatiate on the fact that sex is "old hat." In the last chapter we have seen how the male-female machinery for sexual reproduction was perfected in the ocean by the two-animal animals. Ocean larvae are prototypes of human embryos in the sense that larvae are not accomplished individuals, they are unable to breed a next generation of larvae.† A larva is merely a transitional, nonsexual being, stalling for time to de-

---

* Wolfgang von Buddenbrock, *The Love-Life of Animals* (London: F. Muller, Ltd., 1956).

† There has been exciting speculation, though, that in some cases precocious larvae may themselves have put their sex organs to use, producing new lines of "adult" larval beings, which could lead evolution off in dramatically different directions.

velop generative organs which it will hand over to a bigger and different animal.

A marvelous aspect of metazoan multiplying by sexual reproduction is the way the shape and size of every species of animal is designed for one identical act—the uniting of male and female chromosomes. So it is with oysters, so it is with eagles, so it is with your dog and with you. Sex is no special treat of the human race. If any species can claim proprietorship of sexual reproduction it would be the ocean larvae which evolved the male-female equation.

Each structure, organ, and protuberance, plus each specific characteristic such as size, color, behavior, was represented by its personal gene. These were lined up side by side in a precise order on a flat tape. (In this day of tape recorders this does not seem too miraculous.) Then, a way had to be found to unite pairs of genes with the same *characteristics* from two individuals. This called for some precise doing, and, in that ancient ocean world, some unprecedented new body equipment.

As we have seen, the larvae stage of the two-animal animal allowed time for this fussy reproduction equipment to mature. Concurrently with the development of the generative organs inside the larvae, a strange lottery is played by X and Y chromosomes. They are tossed back and forth with the result that an utterly different sort of "two-animal animal" is created—a "plural animal" which instituted the Adam-and-Eve mode of metazoan reproduction.*

---

* Here is how the X-Y lottery is played in us a thousand millennia after it was invented in the ocean. Every cell in the human body has a nucleus composed of 23 *pairs* of chromosomes. (A chromosome is a collection of genes with similar characteristics.) Thus, a total of 46 chromosomes holds the human code. Each 23 was contributed by a parent at the moment of sexual union. The two tapes adhere their full length, with corresponding male and female chromosomes joined. *All* female chromosomes are paired precisely. So are the male chromosomes, with one odd exception, termed *the X-Y chromosomes*, located at the ends of their tapes. Men always have an X paired with a Y, women have two X chromosomes.

In sex cells the pairs of chromosomes have split apart, so sex cells contain 23 chromosomes, half the standard number for a human being. The full count is restored in mating. Then either a male Y or a male X joins with a female X. Every female human being is XX, every

How could tiny, half-baked larvae, unable to perform sexual intercourse themselves, play a vital role in the evolution of such rituals as exist today?

This question was partially answered in the last chapter (if we accept the Garstang theory), where we saw how in ocean time the larva was the site of the basic evolution of species, more important than mutating gene codes in adults of recent epochs.

The larvae vibrate and gyrate and run with the currents—now in surface whitecaps, now in purple water underneath. The most primitive types are as plastic as drops of water, and they are borne hither and yon in turbulence, their transparent bodies pierced by radiations at all angles and exposed to the slightest changes of temperature, bathed in solutions of organic energy which grew steadily more vital as ocean hordes perished. This was the ideal situation in which larvae perfected the bewitching phenomenon we call *sex*.

The crackling little word stole innocently into our language from across the Channel as Norman *sexe*, meaning "divided." The French had it from Latin *sexus*, a "division." The earliest known mention of the word in writing is in Genesis VI, verse 19 of John Wycliff's manuscript bible which he translated from Latin into vernacular Anglo-Saxon in 1382. The Lord said to Noah: "Of alle thingis hauynge sowle of ony flehs, two thow shalt brynge into the ark, that maal sex and femaal lyuen with thee."

There is no record of the word in the first printed "paper-bounds" put out in naughty London and called "violent, surreptitious, horror" pamphlets around 1450. It first appeared in print in a religious tract, Sir Thomas More, 1532: "I had as leue he bare them both a bare cheryte as with the frayle feminyne sexe fall to fore in loue." Shakespeare, who did not overlook a single beguiling word of his day, mentioned the word sex only once. In *Julius Caesar*, Cato's daughter who is also the wife of Brutus, says: "Think you I am no stronger than my sex, being so fathered and so husbanded."

---

male is XY. The chance is 50-50. In the long run this produces equal numbers of males and females.

The word did not become exciting until our time when mass media discovered that the one-syllable word "says everything" with the greatest economy of printer's ink and time on the air.

SEXUALITY IN THE PRIMORDIAL SEA

Primitive larvae developed the Adam-and-Eve tradition, but *sexuality* is a universal quality of the living cell, linked to an inherent impulse to multiply. Actually, the only kind of life in which no trace of sexuality has been detected up to now are the blue-green algae, which break in two to multiply with utter independence.

We noted earlier the Pandorina, composed of only eight cells, which represents an inchoate stage in the evolution of many-celled organisms. To multiply, it must dismantle itself and each single cell perform fission that, with good luck, produces seven more Pandorinas. In so doing it is forever reverting to the one-celled state. Without sexual reproduction this being was merely a flash in the pan. But before writing off Pandorina as a dead end, let's see what happened.

A larger form of Pandorina turned up, named *Pleodorina*, with a few hundred cells, some of which are a different size and shape than the others. These special cells unite in pairs, and each pair grows into a full-fledged Pleodorina. They are *reproductive cells*, which may tell us how sexual reproduction occurred "a billion" years ago in the ocean.

The scientist who named Pandorina and Pleodorina must have had a wry sense for nomenclature. You will recall that Pandora was a seductive girl whom Zeus sent down to earth to open a box of trouble for the human race. "Pleodorina" is derived from the Greek word *pleo*, meaning "full"—like a "pregnant little girl."

Bacteria, the contemporaries of blue-green algae, have long been a prime exhibit of multiplying by fission. It was dictum that bacteria do not have sex because they have no nucleus with chromosomes. The electron microscope recently perceived DNA floating around in the ectoplasm of one strain of bacteria.

128

The exciting discovery was that these loose DNA's held genes arranged in a definite linear order.

This particular strain almost always multiplied by simply breaking in two. But when tireless researchers pursued these odd bacteria doggedly through thousands of generations, at last they saw two bacteria lie side by side and *conjugate* for half an hour, during which some genetic material slowly passed from one to the other.

Note that conjugation, unlike sexual reproduction, does not join two sets of chromosomes to form a new individual. Conjugating bacteria do not become parents. Afterwards the two individuals go their way and proceed to multiply by fission. But conjugating—which probably occurred soon after the appearance of life on earth—is a gesture which led to sexual reproduction. It came about when the bacteria's vitality was drained through fission multiplying. When one (the donor, euphuistically called "male" by biologists) transferred a portion of its genes to the recipient ("female"), the latter acquired fresh characteristics giving it a better chance to survive and propagate conjugating kinds of bacteria.

Today's laboratories find sexuality on every hand. The late Dr. Ellsworth C. Dougherty, distinguished microbiologist, regarded sexuality as "a very primitive property of living systems, possibly of a single origin in the evolution of life." The evidence for this is that DNA, the elaborate molecule of inheritance, is exactly the same phenomenon in all organisms—in bacteria, seaweeds, protozoa, larvae, fishes, reptiles, birds, and mammals. Evidently the sex transaction originated with a few radical bacteria on the primordial globe.

Not only bacteria, but also single-celled green algae that blossomed out of the static old blue-greens multiplied by sexual reproduction. Where there is no clear distinction between the two kingdoms in primal things before protozoa, those with green chlorophyll that made sugar for themselves out of sunlight are "plants," and those that live on phosphorus, sulfur, iron, and blue-greens are "animals." Both categories enjoyed sexuality with equal profit as it turned out.

A single-celled plant so downright aboriginal as to be boring, is *Chordatella*. It has no mobility and moves about merely by riding passively in water currents. It is not even listed in standard books about algae, and has been generally ignored as having any interest for microbiologists—except to one persevering girl in the biology department of Carleton College, Northfield, Minnesota. She outlasted the Chordatellas while keeping a sharp eye on them as their vitalities waned. Suddenly, when specimens in the circle of the microscope had grown weak to the point of death, she saw a startling event. The paltry alga, too forlorn to divide, suddenly sprouted exceedingly tiny cells inside its body, each of which had two hairs that it waved to swim around!

With the whole biology department watching excitedly, they grew inside the tired "mother." When full grown they crowded toward the center, the mother cell swelling around the waist with a peculiar pregnancy, and active cells breaking out, lashing this way and that in search of another Chordatella to which to attach themselves. This process is neither fission nor conjugation. The cells are not independent organisms, but are momentary motile cells. The report calls them "sperm cells"—but they existed eras before true sperm.

One-celled diatom plants have surprises for those who patiently study them also. Nothing would seem more unpromising for sexual reproduction than an organism enclosed in a glass box. Yet those sparkling one-celled plants in their filigreed glass boxes show another way that sex began in the ocean. The smallest diatoms are transparent and hard to detect with ordinary microscopes, so the vast ocean hid this secret until electron microscopes spotted it.

Diatoms, you recall, get smaller and smaller by fission (see page 70). When one reaches critical smallness, so that a glass box the next size smaller could not hold necessary diatom vitality, the half-box dispenses with glass and hovers naked in the sea.

Quickly other naked diatoms in the same predicament find each other and cling together in jelly blobs while they turn into *gametes* (male and female cells). These mate to create fresh, able-bodied diatoms which soon grow to standard size and enclose themselves in glass boxes.

Ten years after the sex life of diatoms was announced, these "pastures of the sea" divulged more thrilling news about the beginnings of sex in the ocean.*

Naked diatoms scarcely distinguishable from water are an ephemeral living stuff, quite fluid, and a remarkable revelation of an extremely vulnerable organism resorting to sexual reproduction to survive. This is not an individual with sex organs which produce male and female gametes—rather the whole thing, minus its glass box, will fragmentize to produce gametes. So what triggers this extraordinary event in an organism that usually multiplies by conventional fission? What makes gametes male or female? There was no X-Y roulette at this primitive stage of sex.

Prying into the sexuality of diatoms has led to a fascinating field of research for today's biology, namely, the effects of the environment on sexual reproduction. By going deep into ocean time, long before larvae were making gonads for metazoans from established inheritance codes, it is found that *overcrowding* changes the chemistry of the water and this induces sex in diatoms. It was not just smallness, but rather the physical fact that with a particular small size they crowd closer together. The length of daylight (the number of hours the sun is above the horizon) also exerts an influence. And the *brightness* of the sunlight determines whether the diatoms's gametes are more apt to be male or female. In bright light only male sex cells are produced, in dim light only female.

In the case of overcrowding, sexual reproduction seems to be a desperate resort in a crisis. When there is not enough food to go around, the diatoms crumble—tiny gamete cells have a better chance to survive. As for the chemistry of the water, it has been discovered that manganese must be dissolved in the water for diatom sexual reproduction. This metal is termed "the fertility factor." Vitamin B_{12} and calcium must also be available for diatoms to reproduce in this unaccustomed way. And diatoms are much more apt to resort to sexual reproduction where daylight is sixteen hours or longer.

It is arresting to discover that a single-celled organism which multiplies with commonplace fission is able to use sex in a pinch.

---

* *BioScience*, April, 1965.

Early metazoans seem also to have dabbled in this creative technique.

The usual sponge method of reproduction is called "budding." A cell swells like a bubble, breaks away, and starts a new sponge when and if it hits a firm spot. Without the advantages in budding of mixing genes, sponges are a dead end of evolution—but they do show us sex organs trying to develop.

Cells anywhere along the canal tubes of the sponge's body may be inspired to transform themselves into male sperm. They take the shape of little bags which quickly fill up with sperm, suggesting rudimentary testicles. Ripe sperm is discharged into the sponge's canal labyrinth, to be expelled into the outer world through an *osculum* (one of the bigger holes of a sponge). With luck in haphazard currents they may be sucked into an osculum of another sponge, where they copulate with egg cells lined up along the walls of the second sponge—those expectant cells are mysteriously ripe and ready for the arrival of the sperm. This synchronization is doubtless a chemical control. The egg cells shaped like funnels are wide open and ready, with a little whipping hair that protrudes from the mouth of the funnel and sets up an eddy that draws in the sperm. However, shapeless sponges never succeeded with separate male-female sex at all.

They are *hermaphrodites* (a beautiful word from Hermes, son of Zeus, and Aphrodite, goddess of love, thus meaning "male-female"). Hermaphrodites are individuals which take both roles, male and female. In some hermaphrodites, as Professor Berrill * points out, "the sexes have become separate nevertheless, although it is a separation in time and not between different individuals."

Look at an oyster or a shrimp. Is it a "he" or a "she"? Only an expert can tell, with the help of a magnifying glass. Individuals are all about the same size and shape, and at any given moment one might be either him or her. The same individual produces sperm for a while, then eggs for a while, and back to sperm on a timetable that puts the two sexes out of phase and

* Norman J. Berrill, *Sex and the Nature of Things* (New York: Dodd, Mead, 1953).

thus safeguards these creatures from self-fertilizing with the limiting effect of inbreeding.

The Professor, transposing this to human beings, fantasizes that "men would be half . . . the age of women, with the prospect of maternity always ahead of them. . . . No presumptuous male dictator would have the chance to throw his weight around, and great would be the wisdom of old age."

Hermaphrodites seem to show us a stage in the evolution of sex halfway between the organisms that gush clouds of eggs and clouds of sperm into the ocean water to collide and unite haphazardly—and the later, elaborate mates which seek partners of the opposite sex.

DID OUR SEX COME FROM BROWN SEAWEEDS?

Two-sex reproduction was in full swing with the larvae. The crux of it is the mating of two sets of chromosomes so that genes representing the same characteristics from each parent will be paired. Ultimately the evolution of gonads set in motion so beautifully by the larvae would proceed without a hitch to serve land animals such as a mouse, and an elephant, which is a "million times heavier than a mouse."

One vital arrangement had to be made at the very beginning of bisexual reproduction. Mathematics is a law of nature which cannot be violated. It takes 46 chromosomes to make a human being. Now if that number in a male is added to the same number in a female by mating, the offspring would have 92 chromosomes, and these again would be doubled by the next generation. An impossible situation!

At some time, a phenomenon called *meiosis* occurred and solved this problem. Meiosis is a kind of cell division which happens only in sex cells, by which their number of chromosomes is halved—to 23 chromosomes in human sex cells. All other cells in the body retain the full count. Thus, in mating the full count is restored by uniting the half sets of each parent.

This is so universal and so automatic that one tends to take

133

this incident for granted. Yet meiosis is the *sine qua non* of bisexual reproduction. The watery organisms that evolved this device will be forever unknown. We only know that it eventuated with the aid of the law of survival by natural selection even before the era of the larvae. An up-to-date diagram of The New Tree of Life, has something interesting to say on this subject.*

This tree depicts "the single-kingdom" system, with one mainstream of evolution, that of the Plant Kingdom. The original forms of life were algae and bacteria, which are classified as plants. All other living things, including members of the Animal Kingdom, arose from the plants. In the matter of nourishment, animals are entirely dependent on plants. The New Tree of Life is well conceived in the light of today's microbiology—the universality of protein and the identity of the elaborate DNA molecule are indisputable evidence of a single mainstream of organic evolution.

Along the trunk of this new tree are points where things like ferns, mosses, and fungi branched off. They became practically dead ends. Near the summit of the tree trunk as it mounts through ocean time, and just before it reaches geologic time, there is a highly exciting fork which marks the point where bisexual reproduction came into play. According to available evidence, the first many-celled organisms to use male and female gametes, which they set free to find each other and to mate in the ocean, were the *brown seaweeds!*

Brown seaweeds are a distinctive division of the Plant Kingdom, the *Phaeophyta*. They are conspicuously successful. The class includes *fucus*, that long flat shiny seaweed torn from undersea and tossed up on the beach, called "witch's apron strings." The giant kelps in the Pacific Ocean are brown seaweeds. They grow on the continental shelf and extend stalks hundreds of feet long to reach the sunlight at the surface of the sea. (An official measurement is 330 feet, which surpasses the towering redwood trees and makes this the longest, tallest living thing on earth.)

The fact that bisexual mating first succeeded, presumably,

---

* Lawrence D. Dillon, *The Science of Life* (New York: The Macmillan Company, 1964).

with the brown seaweeds, and that this uniting of characteristics gave them such strength and endurance that they evolved so superbly, is a matter of keen interest to us. At that same point on the trunk of this Tree of Life from which the brown seaweeds branch, we see another branch—that of the metazoan animals, the *Subkingdom of the Animalia.*

Since the brown seaweeds long preceded human beings on the planet, we may salute them as our sexual ancestors.

135

12

Love and War Under the Waves

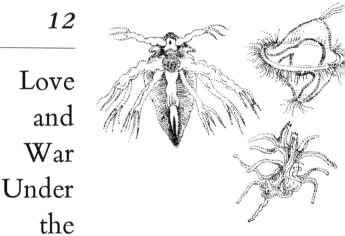

The ocean currents are beginning to look more familiar. They are visibly coming alive. Bigger living things with waving tails and vibrating legs are darting around in the passive plankton. The hunt for mates is on, never to pause to this day.

We glimpse a misty era when crossbreeding has been established as a spectacular success, and the phyla that will lead to our living world are in working order. The first war clouds are gathering over metazoans as their populations fight for more room in the ocean, and contests between predators and prey grow fiercer. At the same time sex is booming and developing incredible resources of mating, using highly imaginative devices.

The scene has a spooky resemblance to today's wars in the air with missiles and supersonic airplanes. Both are fought in three-dimensional space with unthinkable equipment, and what it takes to survive must be created with the utmost speed and ingenuity. Moreover, the evolution of equipment for sexual reproduction obeyed the law of natural selection for survival, and so also must electronic devices obey the laws of existence. To put a man on the moon is a thrilling stunt, but he must take along his water, air, and food.

Trial models with weird ways of mating must have been eliminated in astronomical numbers by natural selection. But the vanished beings could not have been more preposterous than some living in the ocean today. They are lively evidence of the creative genius and power of sexual reproduction. We have learned how the freshening and strengthening of gene codes through mating made living things extra speedy and strong, for attack and escape. Or perhaps sex permitted them to remain almost invisibly miniature while shaped to survive by hiding in tiny niches; perhaps it gave them extraordinary durability. So long as gene pairing helped survival in the wars in the ocean, there were no limits to its inventiveness.

Ears and eyes, fins and feathers, bones and muscles, tree trunks and flowers are means of promoting survival—but for individuals *only until mating and the raising of the young*. Thereafter, species survival shifts to a fresh pair of copulating couples. This process gives an aspect of immortality to gene codes—at least endurance enough for life to evolve along organized lines of phyla, families, and species.

Apparently the chances of death before mating have greatly diminished since the era of ocean time when sex was new—at least in the case of robust land animals that measure their lives in years. These need produce only more than two offspring to promote the species.* But what a *long chance* it was in the era when larvae were originally trying out and developing the male-female equation! The ocean was a treacherous cradle for metazoans. They fought perpetual disruption in the rough-and-tumble situation with ever more complicated bodies and equipment for pairing chromosomes in a hostile world.

At our level of evolution—the outcome of the success of the larvae—the human predicament echoes the crises in the ancient ocean. The more effective the equipment of offense and defense of a species of metazoans, the more effective must be the analogous equipment spelled out in the gene codes of competing species. Thus, through the mighty law of survival by natural selection, the phyla extended across the horizons of millennia.

---

* With the usurpation of the globe by man, the chances of over-topping death have diminished alarmingly for all kinds of life—as a result of spoiling the environment, unnatural speeds of locomotion, and various kinds of explosions.

But there is a basic difference between then and now. *They* had tremendousness of ocean and infinite variety of niches and a thousand million years in which to breed.

THE SEX SECRET OF SUCCESS

Looking at it as a practical matter, there is no logic to us and our living world. Life on earth should have been microscopic forever.* By merely dividing their bodies into two equal parts, without anybody dying, the single-cellers established a 2-to-1 ratio as a sure-fire success.

But this fast, simple multiplying by geometric progression could not be used by big, elaborate individuals with complex organs. For this the ratio of 2-to-1 is inadequate because the chances of death are enormously increased. Giant adults cannot be produced in clouds, and besides, they are doomed to be always at war with each other because they depend on a predator-prey arrangement for food. Now the adult must die; it is the species that is to survive.

The wonderful world of fish, mammals, and birds, of seaweeds, trees, and flowers—using the male-female plan to preserve gene codes—had to have a prerequisite to meet a precarious situation, in which the chance of death of a big individual was enormously increased and survival was a long chance indeed. The prerequisite is the miracle. Instead of a 2-to-1 ratio it became a matter of countless multitudes versus one to survive. The male and female *gametes* (egg cells and sperm cells) of bisexual reproducing reflect the countless hordes of one-cellers. At this critical point we behold the elaborate metazoa imitating the clouds of one-cell ancestors in the ancient ocean.

Moreover, they retain the nourishing, liquid environment of the ocean to carry out their assignment. Taking the human animal as our example, several hundred million sperm at every

---

* It came close to being "forever." The blue-green fossil in Rhodesia, oldest known life on earth, is dated 3.2 billion years ago. The first recognizable metazoa inaugurated geologic time 600 million years ago. One-cellers must have been the sole inhabitants of the planet for some 2.5 billion years.

ejaculation are launched on their journey in the semenal stream toward a highly unlikely chance meeting with a single fully ripened egg cell. When that chance meeting occurs, the egg, electrified, starts on its course in the oviduct river to live nine months in the amnion sea.*

Gametes are not beings, as are one-cellers. They do not use fission to reproduce or generate themselves at all. They are only counters in the face of tremendous odds against the metazoa in their great gamble of survival versus death. In the ocean, where this all started, a beneficial result was that perishing gametes fertilized the water with a perfect combination of vital elements.

So sexual reproduction boomed. Metazoans became more and more elaborate. The production of gametes per copulating couple soared. And thus, in dateless ocean time, the phyla of our plants and animals streamed like shadowy wraiths in the roiling sea.

Today we behold a single corn plant gushing 50 million pollen grains. The pollen produced by a single big sequoia tree forms a yellow cloud against a blue sky above the tree. Mammals, with internal transfer of male gametes, do not require such extravagance as windblown pollen. Yet the count, for the human animal at least, is sufficient for success. As Professor Berrill wryly puts it: "That single egg descending from an ovary inexorably in every woman once a month throughout a third of a century is as much a threat to the peace of the world as the earth has ever known."

Female gametes have a much lower ratio than male in all animals, because they wait passively for the arrival of male gametes eagerly swimming after mates. Yet a human female is born with many thousands of immature egg cells dormant within her. This count is reduced by the menstrual cycle all through her child-bearing years.

The underwater parallel is a female cod laying 6 million eggs free in the ocean. A female salmon in its breeding place far inland at the headwaters of a river ejects 30 million. An oyster,

* Norman J. Berrill, *The Person in the Womb* (New York: Dodd, Mead & Co., 1968).

swinging up the lid of its shell, throws away half a billion eggs per breeding cycle.

The record count for eggs of an ocean metazoan was made by the late Professor MacGinity at the Scripps Institution of Oceanography, La Jolla, California. With incredible patience he tallied the eggs of the sea hare, which comes out of the deep to spread tangled threads of salmon-colored egg cells in the eel grass of an estuary—478 million eggs deposited by one mother steadily for eighteen weeks, at an average rate of 2,640 per minute.*

This sample shows how sexual reproduction protects the integrity of species with astronomical numbers of gene codes at the time when the chances of death are astronomical. Thus, male-female capers produced fantastic menageries—and will continue to do so as long as there is a healthy ocean to work in. When life emerged upon the continents, individuals brought along the Adam-and-Eve plan, with gamete stages maintained in various facsimiles of sea water.

Trees are an arresting exhibit. Their sap streams deliver high into the air dynamic solutions of iron, magnesium, phosphorus, sulfur, etc., like those in the ocean. When buds containing the reproductive organs are readied for business (usually in the spring), the arms of the ocean in the form of threads creep upward against gravity, quicken the flow, head for the pinpoint targets at the tips of twigs, and bathe the male and female organs. Then pollen foams from anthers, and ova at the bases of pistils swell.

All land animals—earthworms, houseflies, snakes, butterflies, birds, mammals, name what you will—must unite their gametes in a watery environment, an "arm of the sea." They must somehow swim together in a fluid,† with the same function as sea

---

*Lorus J. and Margery Milne, *The Mating Instinct* (Boston: Little, Brown, 1954).

† Today the word "humor" refers to any of the major animal fluids. Hippocrates, sitting under a fig tree on the Greek island of Cos in 360 B.C., expounded four humors of the human body—blood, phlegm, yellow bile, black bile. So we call people sanguine, phlegmatic, choleric, or melancholy. Benjamin F. Miller and Ruth Goode, *Man and His Body* (New York: Simon and Schuster, Inc., 1960).

water, namely, the meeting place and mixer for sexual reproduction. Moreover, in every animal body the gamete mechanisms
are secured in sanctuaries, like potent low-tide pools.

THE SOVEREIGN MOLECULE OF LIFE

If your imagination seeks one common ancestor for the Tree
of Life, it should be a DNA molecule. This chemical, of which
chromosomes are made, is universal in life. It is the reigning
molecule that holds the key to every species. Never found apart
from the living cell, DNA must have first turned up in primordial bacteria and blue-green algae. Today, molecular biologists are pursuing the electronic trails of protein and its role
in DNA back into the depths of the timeless ocean.

As this is written, a report comes describing a provocative
experiment in which they isolated *histone*—a kind of protein
which plays a key role in the behavior of DNA by acting as a
brake on genes, allowing only selected genes to do a particular
job. Does this help to solve the profound puzzle as to what
makes the similar genes produce different organs, and DNA
create things as contradictory as an eagle and a mosquito, or an
oyster and a scientist? This discovery seems to indicate that
histones can recognize different genes!

As usual with explorations of the universe inside living cells,
discoveries of "facts" bring deeper mysteries.

The clipping on my desk says that histones were isolated
from the flower buds of peas, which contain the sexual apparatus
soon to bloom, and also from the thymus gland of calves. The
thymus is thought to be a very primitive gland of the mammal
body. Its hormone is called the Peter Pan hormone because
when it is injected into the pupa of a moth (the resting stage
between the larva and adult) it delays the pupa's development
as a mature moth, by putting the brake on certain genes.

This has caused some excitement, as expressed by Miller and
Goode: * "We are tempted to speculate that the Peter Pan

---

* Benjamin F. Miller and Ruth Goode, *Man and His Body* (New
York: Simon and Schuster, 1960).

hormone may have some youth-prolonging value to human beings which could postpone the aging process; but investigators are cautious."

When single histones from the pea buds and the calves were compared, their structure was discovered to be "remarkably similar." The report concludes: "Apparently the composition of these histones has changed little since the pea and the calf evolved from a common ancestor."

And, let me add, human histones are precisely the same chemical as that which controls the sexual reproduction of peas and calves—and it came to us all out of the ocean.

Though that glimpse of a key molecule in the internal ocean of our bodies can be perceived only by scientists in the inner sanctum of laboratories, human gametes are visible through an optical microscope in a doctor's office. The sperm cells resemble a bevy of tadpoles swimming vigorously in their sea, a droplet of mucus. The head carries the code of the father's characteristics, and this is driven forward by an undulating tail—which detaches when a sperm hits its target, an egg cell.

In copulation a sperm has a journey of about 10 inches through the uterus into the canal of the Fallopian tube. This is quite a journey for a "pollywog" only 1/500 inch long; at the rate of 1/8 inch per minute it takes approximately four hours to reach the ovum, the female gamete.

The sperm revolves, corkscrewing its way with a helical wave—the dynamics of a ship's screw. It yaws and rolls as it fights to move upstream against the current. (In a later chapter we shall see a slender baby eel swimming upstream in similar fashion.) The head-on current is caused by waving hairs (cilia) which line the wall of the Fallopian canal, now pushing the otherwise immobile ovum down toward the uterus to meet the sperm.

Recently, in probing the meiosis (sex cell division) that precedes the actual act of reproduction, it was found that the cells that will be sperms and ova move into position by poking out protuberances and withdrawing them like amebas, thus mimicking the motions of those very primitive single-cellers.

After this little marine drama of human gametes is accomplished and an embryo is bred, the forming baby girl or boy will persist in looking and acting like an underwater animal—which, indeed, the unborn human baby is with its fishlike tail and gills. Not until the seventh week after pregnancy does this embryo show human characteristics.

It continues to live a submarine life in its own inland sea, the amnion "bag of waters," in which it floats buoyant and carefree for those nine months while the home sea expands with the growing embryo. When everything is ready, the water breaks and rushes out, the lungs expand and take over breathing from the gills, "there is a dramatic revision in the functioning of the heart" (in the words of a medical man), and the baby utters a cry to proclaim its emergence from the sea.

THE HYDRA TELLS A WEIRD TALE

Sexual reproduction pullulating in ocean currents, through infinite time with boundless space to maneuver in, developed some outlandish beings—from our viewpoint.* We look far back into ocean time, when sex was trying all kinds and sizes of body designs, and find some of those odd models still alive to tell their tales.

What sense does the hydra make? That such a thing can now be living on earth is worth pondering. About half an inch high, hydra has poison-dart † studded tentacles that can reach out in all directions to grab food. This gives it a definite shape, like a bit of string frayed at one end. By shuffling along—like a lazy

---

* Elsewhere I have written about our myopic outlook: we see all other creatures as odd or funny, with the exception of some dogs, horses, and pussy cats that seem to share our admiration of ourselves . . . yet the giraffe glancing down from his stepladder would doubtless think that a man is very funny with such a short neck. A chickadee would be justified in its opinion that a man is ridiculous in the way he struts around, and would get a laugh out of man's contraptions for flying. What would a walking-stick think of buttocks and breasts? Isn't it possible that an aphid (which can produce fifty babies at a clip with no help from a male) would regard two people kissing as ridiculous?

† The famous nematocysts, which have awed biology students.

fisherman who moves by bumping along on his buttocks—it can move a short distance in search of a better place to fish. Driven by hunger to really move, hydra shortens its tentacles, bends double, and using them as a tripod it heaves itself up and over with a cartwheel like a tumbler—it has no upside down, it may hold poised with a beautiful handstand.

Hydra reproduces by budding—new hydras simply sprout out, break off, and go their way. But occasionally hydra indulges in sex by releasing male and female gametes that are attracted to each other, get together, and mate.

Occasional sexual reproduction rescued hydra from being a dead end and put it at the head of a populous phylum which includes jellyfish, Portuguese man-of-war, sea anemone, and the coral animals.

Sex did wonders for hydra but the creature is not just a good show, it tells a tale about why sex operates only in water, and what makes one kind of cell turn into other kinds of cells—the great mystery of *differentiation*.

Here is an animal so primary it can regrow its parts. Cut a hydra in two—it forms a new head on the part with the foot, and a new foot on the part with the head. Or two heads from other hydras can be grafted on and the animal will live happily with three heads. Thus, hydra is named after the Hydra with nine heads that ravaged the country of Argos. Hercules was sent to slay it, but every time he knocked off a head with his club, two new heads grew in its place.

Added to this, the hydra has skin cells that can reproduce hydras by nonsexual budding, but, as we have seen, the same cells may suddenly become male and female with an urge to mate. Here was a perfect setup to investigate the towering mystery of the single egg cell dividing and dividing to create organs, tissues, and bones of big animals—and also to try to shed light on the way males and females shaped up in the pre-larvae sea. The answer is quite sensational and it has pointed the way to some of the most exciting current research into the secrets of life.

Hydras are ideal guinea pigs for these studies of molecular secrets. They stay in one spot most of the time, and they are quite durable and obliging to scientists who submit them to all sorts of indignities. They have been grown in distilled water with salts added, and in tap water, and in water from a very stagnant pond. They have been shaken up with oxygen and carbon dioxide gases, severely chilled in a refrigerator, and fed on extraordinary diets. Their tedious nonsexual life has been charted interminably, and their sexual life recorded to the last molecule. The outcome is the discovery of the "Sex Gas of Hydra"—the title of the report by Dr. W. F. Loomis, who made the original experiments.*

A breakthrough came when it was discovered that stagnant water in an icebox incited sexual reproduction. Next, the hydra watchers found that hydras in a goldfish bowl containing guppies, etc., turned sexual so regularly that this particular bowl was called by the researchers "the sex bowl." Moreover, when hydras became overcrowded in segregated places they reproduced sexually.

Then step-by-step a common denominator of these disparate events was found. They all had a high pressure of carbon dioxide gas. This results when the carbon dioxide exhaled by organisms is not readily dissipated in air currents brushing the surface. With the goldfish bowl this was due to the small opening at the top compared to the large volume of water. In the crowded place the hydras breathed out CO_2 faster than it was lost. A fraction of 1 percent more carbon dioxide increased sexual practice among hydras as much as 40 percent!

Research into the chemical nature of water and its effects on living things is continually making fresh headlines. It is telling today's scientists much that has been utterly unsuspected, from the worms of Aristotle in the third century B.C. to T. H. Morgan's fruit flies at Columbia University in 1940—revelations about the mysterious travels of salmon thousands of miles into Pacific Ocean deeps and the return to their precise birthspot thousands of miles upstream in the Yukon and Columbia rivers. The temperature and chemical solutions of oceans influence long-distance bird migrations, most vividly that of the Arctic

---

* W. F. Loomis, "The Sex Gas of Hydra," *Scientific American* (April, 1959).

tern, which flies annually some 20 thousand miles round trip through fog and storm over beaconless ocean between the Arctic and Antarctic regions. And somehow these conditions control the grunions that suddenly emerge from deep water in the Pacific on the same day every year, to ride up to high-tide mark in the spreading waves on a California beach.

We shall return to the grunions in a later chapter, in which we shall behold what sea water brought forth in the quickening tempo and the surge of living energies that resulted from the triumph of sex. We have reached a dazzling stage of evolution in which the serious business of the reproduction of creatures of our living world arises out of the grand sweep and carefree manner of the ocean.

13

Fantasmagoria

No creation of man's brain is original. His senses borrow every element of imagination from the surroundings. Sir Thomas Browne (early seventeenth century) brilliantly exemplifies the point. He was a curious mixture—a successful physician (by training a scientist), imaginative, with the nature of a philosopher and of a melancholy mystic. William Rose Benét calls his prose "rich in striking images and one of the most remarkable accomplishments in English literature." In his book, *Vulgar Errors*, Sir Thomas points out that the anatomy of the lion and the unicorn on the English royal coat of arms would make it impossible for the unicorn to copulate except backwards . . . and that one is "not able to judge of male or female, or determine the proper sex" of the unicorn.

Around 1600, natural history and folklore were intertwined. Unicorns were as real as antelopes. A hunter stated that "Unicorns may be captured by making them run their horns into trees, behind which the huntsman has dodged." Even Shakespeare says, in *Julius Caesar*, "He loves to hear that unicorns may be betrayed with trees."

Later when life sciences became more sophisticated with the great naturalists—such as Linnaeus, Buffon, and Cuvier—the unicorn was relegated to mythology. This brings me to a peculiar example of how the phyla created in the ocean produced beings far more "imaginative" than anything a human brain can think up.

Let us put the unicorn together again—this time not in a bestiary but in the ocean. I have seen the animal with my own eyes in its hideout in a fjord in north Greenland, vicinity of Kane Basin. It has a fat body of around 500 pounds, no limbs or well-defined head, but out of a mouth at the front end of the male projects a spiraling, ivory javelin, perhaps 9 feet long.

So beautiful is this spear that the Vikings who hunted Arctic waters around 1,000 A.D. brought them back to Europe as gifts to chieftains. The traditional throne of the Danish king is supported on sea-unicorn horns. Who then could have doubted the reality of unicorns? Hearsay, without written records, could put them in Africa or in forests along the Rhine, cohabiting with antelopes and lions.

Strangely, there is convincing evidence that ancestors of the sea unicorn did in fact have limbs and prance on land. The animal is officially called *narwhal*, relative of whales and dolphins. They are warm-blooded and air-breathing, and suckle their young like other mammals. Front legs have turned into a pair of little fins, hind legs have been absorbed and left remnants of the limb bones buried deep inside the body. The Milnes in *The Biotic World and Man* catch a glimpse of these animals "living an amphibious life at the seashore, apparently escaping to the open water when pursued on land" some 50 million years ago.*

Absurd as the rigid, clumsy protuberance looks on an animal that should have the grace and agility of a dolphin, we can be sure that it evolved according to the law of survival by natural selection. It is not analogous to the stubby horn of the great Indian rhinoceros (*rhinos*, Greek for nose, *keras*, a horn). That monster's "horn" is not a bone projection from the skull, but a clump of hair stuck together with adhesive.

What possible contribution can this tremendous tooth make to the well-being of the animal? It cannot tear or grind food

---

* Lorus J. and Margery J. Milne, *The Biotic World and Man*, 2nd edition (Englewood Cliffs, N.J.: Prentice-Hall, Inc., 1958).

like a standard tooth. So far as is known, it is never used as a weapon. The animal in its refuge among the polar ice pans with its relatives, the seals, whales, and dolphins, has no enemy. It does not brandish the ivory like a ghostly white sword to protect its feeding ground.

One answer came from the late Admiral Donald B. Mac-Millan, who, in a lifetime of Arctic exploring, confronted more live narwhals than anybody else. The contents of the animal's stomach show that its chief food is halibut, a flat fish which hides by lying still on the bottom. The narwhal impales the immobile prey on its spear point, and then thrusts forward, swimming leisurely while the pressure of the water against the flat body of the fish makes it whirl around on the spiral ridges as it travels the length of the spear to pop into the animal's mouth. How else could a sea unicorn, unable to put its face within eight feet of the fish because of that rigid, nonretractable spear, get the fish into its mouth?

If true life in the ocean can surpass mythological beliefs, what about purely imaginative fairy-tale characters?

Two famous tales that boldly juggle reality are *Arabian Nights Entertainments* and *Alice in Wonderland*. For example, the first has an enormous bird called a roc, which eats a full-grown elephant for strength to bear Sinbad the Sailor out of the jaws of death in the Valley of Diamonds; in *Alice* the little girl can become a giantess or pygmy by nibbling opposite sides of a mushroom. Yet these weird events are no more ludicrous than true life beings and their resources for mating that evolved according to laws of nature in the boom of sexual reproduction following the two-animal animals in the ancient sea.

The sea horse is as quaint a cartoon as ever appeared in salt water. Its eloquent scientific name is *Hippocampus kudo* (Greek for "glorious horse sea animal"). It swims erect, looking like a knight on a chess board. It is a true fish remodeled with tail fin that spirals to grasp seaweeds, and instead of moving by flexing the body sidewise, like typical fish, it jogs vertically to send currents of water up along its dorsal fins, and thus moves about with its body held erect. This unique posture is dictated by sea horse mating procedure, by which the males relieve the females of all their usual functions except the original production of eggs.

The pelvic fins are folded together to form an ample pouch on the abdomen of the male. His courtship consists of displaying the empty pouch as he goes from female to female pleading to each with clicking sounds that she "drop her coins into his tin cup." He is restless, and doesn't pause in his begging, until he finds a female sea horse who is lured to drop her eggs into his pouch. When it is full with about 200 eggs, he seals them in with glue so sea water cannot touch them and swims around looking pregnant while he nourishes the embryos with his own bloodstream. This paternal brooding of eggs is so efficient that sea horses skip the larval stage, and complete sea colts burst from their father's pouch.

THE GHOSTS IN THE OCEAN

A parson was leading a party of boys along the edge of the sea near Cape Otway, Australia, looking for a good spot to fish. Standing on a rock about three feet above the water, he was starting to tell them to put in their lines when he gave an awful yell. Three tentacles of an octopus, each as thick as a man's arm, had slid up the rock and swiftly coiled around each leg and an arm. Five other tentacles, anchored by hundreds of suction cups, began to contract and pull the parson down into the water. It took the strength of six boys and the man to resist the terrible tug and save the parson's life. This incident was recorded in the biological record because an octopus rarely reaches out of the water to seize its victims.*

The adhesive power of the octopus's suction cups is famous —240 of them on a tentacle, 1,920 per octopus. A pearl diver on the bottom, more than a hundred feet below the surface, was suddenly enwrapped by the horrible tentacles. He pulled on

---

* My chief sources for this section are two superb works of research: Frank W. Lane, *The Kingdom of the Octopus* (New York: Hill and Wang, 1968)—and Bernard Heuvelmans, *In the Wake of the Sea-Serpents* (London: Jarrolds, 1957).

the signal rope to haul up fast. The haul-up man couldn't budge him. Then three men tried. The diver said he had the sensation of being pulled in two. It took more than human muscle to save his life. They wrapped the haul-up rope around a stanchion on deck, and as the lugger dropped in a trough they hauled the rope taut with all their strength. When the lugger rose on a swell, the octopus's suction cups at last gave way, unable to resist the power of an ocean wave—up came the pearl diver, still enwrapped by monstrous tentacles when he arrived on deck.

Let me add one more item to the true life adventures with octopuses. A diver of the Royal Navy was sent down to find a practice torpedo lost 40 feet under the surface off Gibraltar. He found a dump from the naval station—old pots, pans, scrap iron —and a suit of overalls which bulged a bit as though it contained a body. He grasped a sleeve—the overalls reared up and swayed before his eyes as though possessed by a ghost! Then something squeezed his ankle and gave it a terrific tug; an octopus had appropriated the denim for a lair. The diver drew his knife, tried to sever the tentacle coiled around his ankle. As he was hauled up, he saw the suit of overalls whirl away and disappear in a black cloud of ink which the monster shot out of its siphon.

Fact and fable are solidly merged in the case of the octopus and its relatives. Although much that we have considered in these chapters has been discovered by the techniques of modern science, not so with the octopus. This ghost haunted the Mediterranean in ancient times and it was a subject of great interest to Aristotle, the father of zoology, who described the anatomy of the octopus 2,200 years ago. He actually recorded the astounding male reproductive organ of an octopus, called the *hectocotylus*—which is ancient Greek for "arm of a hundred suckers"; we shall return to that unique thing presently. Aristotle also noted this fine point: that a female octopus cradles her fertilized eggs in sensitive folds of a tentacle, where she shoots jets of water among them to aerate them, vacuum cleans them with her suckers, and manipulates them to keep the eggs from being eaten by their father.

Pliny, the Roman naturalist, wrote around 50 A.D. that he

151

saw a creature with "30-foot arms" climb out of the sea to eat fish left at the water's edge waiting to be salted.

About 950 B.C. in Homer's *Odyssey*, Ulysses was warned about the dangerous channel between Scylla and Charybdis. Heuvelmans, in his book, says that Charybdis "is clearly a whirlpool," and as for Scylla—she is a monster who lurks in a cave yelping and squealing. She has six necks, each with an obscene head. She keeps buried waist-deep and sways to sweep the reefs for quarry which breed in the thundering waves. She never fails to snatch a man with each of her heads from every ship that comes along. It is clear what sea creature inspired Homer's Scylla.

Somehow the great poet Homer heard frightful animal sounds among the thundering waves. The octopus has no sound organ, it is as silent as deep water—but when it is forcibly lifted out of the water, it bawls through its siphon a racket that "sounds like a yelping puppy."

MASTERPIECE OF OCEAN CREATURES

It is a paradox that the *summum bonum* of organic evolution on land is not only the brainiest but also the most fearful form of life. Ask the bison and the eagle, ask the redwoods and wildflowers, what living thing is most fearful on land. Strange as it seems, octopuses are at the top of evolution of true ocean life. If living creatures had never developed backbones and emerged from the sea, there is good reason to suppose that the kingpin beings of the planet would be the fellowship of the octopuses. I say "fellowship" because the octopus type of organism comes in various forms under the scientific heading of *Cephalopod*. The name itself is arresting; in Greek it means "head-foot." It refers to the close kinship of the octopus and the shellfish— the everyday snail, clam, or oyster, whose most prominent protuberance, by which they inch around, is known as a foot. Biologists noted that the octopus's head is in the same position as the foot of the mollusks. Moreover, the octopus has a shell tucked away inside its body, left useless like the human appendix that is also a relic from lower forms of life. The octopus tribal shell is best known as the cuttlebone, which we hang in the cages of canaries as a beak-sharpener.

What specific features enable the phantom octopus to be called a masterpiece of ocean animals (whales, seals, and dolphins are land animals by heritage)?

Here is a creature with all the exquisite lability of water. Yet unlike the primitive jellyfish, which evaporates and disappears when it dries up, an octopus has flesh and form. The octopus animal has bilateral symmetry, separate sexes, a distinct head, and a highly developed nerve system to serve its keen senses. It is an exquisite design for living.

Yet, men looking down from their element on land have always regarded the octopus as an uncanny and horrible monster. Herman Melville has choice words for this sort of thing in *Moby Dick:*

> We now gazed at the most wondrous phenomenon which the secret seas have hitherto revealed to mankind. A vast pulpy mass of a glancing cream color lay floating on the water, innumerable long arms radiating from its center, curling and twisting like a nest of anacondas, as if blindly to clutch at anything within reach.

Yet behold how the architecture of the octopus animal is a magnificent expression of the fluidity of water!

Frank Lane, in *The Kingdom of the Octopus*, recites an incident told by Roy Waldo Miner, formerly a Curator at the American Museum of Natural History. While collecting on a coral reef in Puerto Rico, Dr. Miner captured a small octopus with a 12-inch arm span. He tucked the eight squirming arms into a cigar box, hammered the lid down with tacks, wrapped a strong cord around the box for maximum security, put it on the bottom of his rowboat, and continued collecting. Upon landing, he untied the box and pried open the lid to show off the young octopus. It was empty! Sleight of hand? Not at all —sleight of octopus! Dr. Miner supposed that the power of expansion at the tip of a tentacle enabled the animal to pry open the lid a crack. Then taking a powerful hold with its suckers outside "it pulled its elastic body through the crack by flattening it to the thinness of paper." An idea for a TV show would be to put an octopus that has a 36-inch span in a wire cage that has a half-inch mesh, and see the creature pour through and escape unfazed.

Designed for living?

Their favorite den is a cave in coral or rock, with a very small opening through which they peer as with an eye at a keyhole, and where, if a giant like a dolphin comes nosing around, they can pile up an abatis of crab shells—crabs are favorite octopus food.

I borrow again from Lane. A two-gallon carboy (a big water jug) was dredged up from the bottom of the English Channel and found to be chock-full with a complacent female octopus whose head was the size of a large grapefruit, plus her eight great tentacles, and a big brood of eggs. Evidently she had poured herself through the two-inch neck of the jug to take up residence in this niche while very pregnant.

Designed to survive? Something more is needed.

A masterpiece of evolution in the ancient ocean must be able to fight off a murderous foe. In this matter an octopus developed a really spooky resource, suggesting the myth about a cat with nine lives. It can be dismembered, lose its arms, and, so long as it has a minimum of three arms left, one for anchorage and two for coiling around and squeezing prey, the lost members will regenerate quickly. But wait! If there is no enemy around to excite the octopus—as in a laboratory experiment—all eight of its arms can be cut off and they will slowly regrow from the stubs which encircle the head.

The two "sea tigers"—the ferocious moray and conger eels —may destroy octopuses that they catch peeking out of the hole of a cave. Its stronghold prevents the octopus from brandishing its ghoulish weapons. Most of its arms are pinned to its side, so it can manage to extend but one arm to brandish at the enemy. The moray grasps the single groping octopus arm in its mouth, whirls around until the arm is twisted off. Out through the hole of the cave comes another tentacle arm—the moray eel whirls, this arm comes off. And so one by one, the octopus loses all its arms and only the head is left for the eel's dessert.

But give an octopus room to operate and the story is different. This was observed at the Scripps Institution of Oceanography at La Jolla, California, where a battle between a moray eel and a hungry octopus was staged in a tank for scientific observation. The octopus took the offensive from the start. An eel is agile and swift too. The weird waving and coiling of the octopus's tentacles seemed to hypnotize the moray. It grew less terrible, after much churning it went down for the count, and this time the octopus ate the head of the moray eel for the coup de grace.

OCTOPUS'S MAGNIFICENT RELATIVES

That the "head-foot" creature represents a climax in the evolution of sexually reproducing ocean animals is gloriously confirmed by certain octopus relatives that are some of the most fantastic and beautiful of all animal life on earth.

The *chambered* (or *pearly*) *nautilus* is the result of enclosing the body of an octopus, which possesses soft fluent qualities from the wild and free ocean, inside a rigid shell. The animal, born from an egg on the bottom, immediately sets about collecting calcium crystals from sediments and selects some particular elements dissolved in the salt water. The nautilus spins and spins its elegant whorls in response to the innate spiraling of water—to describe a geometric figure that is renowned as the most perfect art form in nature, the dynamic, or logarithmic spiral.

D'Arcy Thompson, in his superb book, *Growth and Form*, is fascinated by the nautilus shell. He points out that, although divided into compartments, the gorgeous pinwheel is formed as one continuous tube. An inherent, *disciplined force*, once set in motion by an "ugly" octopus, does not deviate from its mathematical whirling until it has fulfilled the commands of the creature's genes to make a home of precisely the right size.

D'Arcy Thompson observes here a mathematical *law of growth* which defines the nautilus spiral and is common to the ram's horn, a pig's tail, an elephant's tusk—and everywhere you look in the Plant Kingdom, such as the crozier of a fern unfolding in the spring, and pine cones. Water twirling down a drain may be the original expression—or does it all reflect the spiral galaxies in outer space?

Getting back to the chambered nautilus, as the baby grows it repeatedly finds itself overcrowded in successive shell compartments. Then it seals off that place and builds another larger one this side of the partition. So the shell grows in exact proportion to the size of the animal. Finally the fully grown nautilus occupies the outer and largest compartment—often quitting its shell mansion briefly to act like its relative, the true octopus in search of food.

Surprisingly, the graceful shell, which should be cumbersome, gives its owner mobility. Nautilus rides in the ocean currents buoyantly because, when the animal seals off a chamber it fills it with a gas similar to air but with more nitrogen. The bigger the shell, the more buoyant it is.

"The Chambered Nautilus" is the most famous poem by Oliver Wendell Holmes. The last stanza expresses the philosophy with which this beautiful ghost of the sea inspired the poet:

> *Build thee more stately mansions, O my soul,*
> *As the swift seasons roll!*
> *Leave thy low-vaulted past!*
> *Let each new temple, nobler than the last*
> *Shut thee from heaven with a dome more vast,*
> *Till thou at length art free*
> *Leaving thine outgrown shell by life's unresting*
> *sea!*

THE OCTOPUS THAT MASQUERADED AS A FISH

At the bottom of the ocean * the octopus style of body was recast to make a cuttlefish. Instead of being all head and arms,

---

* The Linnaean Society of London reports the dredging up of a cuttlefish from about a *mile* down in the middle of the Indian Ocean —"a curious cuttlefish set with minute sparks of blue phosphorescent light."

it has an elongated body back of the head with slender frills extending the full length on each side. These odd fins have no bones, they merely ripple to make the cuttlefish swim weakly. The eight octopus arms are bunched and reach forward instead of radiating in all directions.

This soft sort of freak octopus was never purged by natural selection, for dramatic reasons. It is endowed with jets at both ends by which it darts forward or backward and also discharges heavy clouds of ink for a "smoke screen." In addition it possesses a fine mechanism for feeding—two sharp tentacle javelins shoot out like lightning to harpoon passing prey, or they can be used like tongs to seize all shapes and sizes of shellfish. With no teeth, it has powerful jaw muscles to crush the strongest shells—even a stony limpet can be yanked from its fastening and crunched to crumbs. A cuttlefish has been observed helping itself to the innards of a dead dogfish by using its tongs like a spoon to convey the food to its mouth.

Indeed, this style of octopus animal is as successful a design for living as the common octopus. Nobody can estimate the cuttlefish population, because they live mostly in dark, mysterious depths.

Cuttlefish are the legendary "fish" of the ancient civilizations of the eastern Mediterranean. In biblical times they were caught by specially designed baskets sunk to the bottom by stones. D'Arcy Thompson, when he was professor of zoology at Dundee University, pointed out that Mideast people catch cuttlefish today in the same way as they did two thousand years ago. Evidence of tremendous unsuspected populations of cuttlefish is revealed when earthquakes disrupt the ocean floor.

Although cuttlefish are not considered fish of northern seas, they or their bodies may be carried far north in submarine currents. In 1944, people in the Shetland Islands were astonished to see cuttlebones cast up in such vast numbers that they "formed a white band along the tide marks." That was a windfall for canaries in cages. The cuttlefish's internal bone is larger than that of the common octopus. It firms the body so cuttlefish cannot swim by flexing horizontally, but must move by its undulating frills and jets. The ink of its smoke screen is a unique

157

concoction of seawater minerals that makes it a beautiful and stable red-brown—the *sepia* of the old masters and artists today.

I feel that these cephalopods show us an epoch in ocean time just before the backboned fishes appeared on the scene.

Does the cuttlefish show us a spooky, watery octopus trying to turn into a fish? The gene code of an ancient octopus mutated, changed in certain octopuses which lived at the bottom and were tempted by the growing snail and clam populations. Most of the gene code was unchanged—the cuttlefish sexual reproduction is the same as that of the regular octopus; it has eight arms, definite head with a brain, and fine big eyes. There is the elongated body of a fish, but no bones—except that big cuttlebone, which may be telling us something very exciting.

The cuttlebone is placed at the top of the back of the cuttlefish—in a position similar to a backbone, although it has no vertebral segments and it contains no nerve cords. The question is being debated by biologists as to whether the cuttlebone is analogous to a vertebrate backbone. Surely its position is analogous, very advantageous for firming and later supporting a vertebrate body.

EQUIPMENT FOR EMERGING FROM THE OCEAN

What we see in the sea and on land are kinds of life whose long lines of ancestors established the phyla with creatures that could stay alive long enough to give their gene codes to descendants. They were the successful ones which had capacities to fight and win over enemies—or else to escape by speed, hiding, or cunning, long enough to reproduce.

Because the octopus fellowship possesses these capacities in the highest degree, I have put them at the top of the ladder of animal evolution in the ocean. The style of body conforms perfectly to the ocean habitat, the creature's sexual routine is superb. Its behavior and organs can be seen and compared to those of human beings. Important among octopus features, as we have noted, are elongated bodies with bilateral symmetry, a head distinct from the rest of the body, a lobed brain that prompts its owner to make decisions—even to pique curiosity

—a blood circulation,* a nerve system, a pair of superb eyes, separate sexes, with highly specialized mating equipment, and maternal instinct for protecting the young. It is not surprising that this "devilfish," alone among the sizable animals of our planet, is unchallenged and ineradicable by man's cunning.

The octopus fellowship is showing us an important way in which organic evolution operated in ocean time. Shapes, sizes, senses, organs, colors, circulatory systems, sinews, bones (shells), flexibilities, mobilities—all such parts and abilities evolved of themselves by adapting to each other. They evolved in the body as well as the ocean environment. Back of that was the evolution of molecules and cells.

It took a lot of different body features to build a mammal in a far distant age. But let me save that subject for the climax of this chapter. First, there are two more fantastic octopuses to be mentioned, which lend exciting evidence to my supposition that the octopus fellowship represents the highest point of evolution for life on our planet, so long as life existed only in the sea.

The little *argonaut* shows us an octopus trying to turn into a bivalve. The others have only internal cuttlebones; argonaut has the cuttlebone and also it conjures up an exterior shell suggestive of a first-rate oyster, pure white, 6 to 8 inches across. The female, when pregnant, raises this gleaming white shield above the blue waves and goes sailing before the wind. When fishermen of the Near East saw these pretty white sails on the Mediterranean, they called them "Argonauts," after the dauntless mariners of Greek legend who were the first to resist the Sirens who inhabited the cliffs of islands in the straits between Sicily and Italy.

The sweetness of their singing bewitched mariners who came within earshot, compelling them to land, only to meet their death. The skeletons of victims were thickly strewn around the caves of the Sirens. No crew ever succeeded in defying their charms, until the Argonauts addressed themselves to this terri-

---

* Octopus blood is blue, as is that of its mollusk relatives, because each molecule of blood is formed around an atom of copper. The red blood of bony fish and land animals is formed around an atom of iron. This does not lessen the originality of the octopus's circulating blood system.

ble problem. Their strategy was to persuade Orpheus to come aboard—he played the lyre and sang so beautifully that fish, birds, and beasts gathered around to listen to him. The Argonauts, with their attention fixed on the sweet music of Orpheus, sailed right past the sinister Sirens.

In this context of love songs and the argonaut's white sail, let us consider an important bequest to future land life—advanced sex equipment, which affords selective, internal mating between individuals (in contrast to the old method of the two sexes expelling clouds of eggs and sperm into the surrouding water to collide by chance). That was a highly efficient way of performing sexual reproduction based on the multiplication table. Oysters, clams, sea worms, sea urchins, and fish still use this impersonal, exterior way of mating. But it could not serve complicated elegant organisms because it is too haphazard to transmit heartbeats and nerve impulses through exquisite networks.

When a male octopus comes of age, *one* of its eight arms—the third right arm—becomes unique and wonderful. It turns into an organ called a hectocotylus (arm of a hundred cups). Now this arm no longer coils with fierce intent to seize a prey, nor does it use its suction cups to anchor to rocks. These hundred cups are soon filled with sperm cells, and the arm gently uncoils to grope and thrust under the mantle of a female and into a pocket where she has collected a batch of egg cells. Fertilization is swift and sure, the eggs awaken as embryos, which the female octopus nurtures under her arm as though in a womb.

The common octopus male withdraws his hectocotylus and goes his way, but in the case of the argonaut and her beautiful white sail, an uncanny episode ensues. This is related by Frank W. Lane.

The hectocotylus breaks away from the male, and since he is very little, what is left of him goes off and dies.* Biologists

---

* This appalling ritual of sexual reproduction in the ancient ocean is reflected today by praying mantises and by spiders, whose males are tiny compared to the females—but in their case the female devours the male after copulation.

studying the argonaut in aquariums to learn their personal secrets have stared in amazement to see the hectocotylus sometimes break away from the male *before* insertion, and go off to swim and crawl around independently as though it had a mind of its own. Early nineteenth century zoologists mistook independent hectocotyluses for some strange and unclassified sea worm. A German argonaut specialist identified the fantastic thing for what it was—a penis—with considerable difficulty because, in his words, "they are very restless, and wind and twist about in the most determined manner."

In her pre-shell life the female argonaut has rarely been seen, as must be true of many queer creatures today in the profound ocean wilderness. It is known that she leads a free swimming life like other octopuses, and she probably lives among rocks near the bottom waiting for a male to come along brandishing his hectocotylus—or perhaps she takes possession of an independent hectocotylus swimming or crawling in the vicinity.

Soon after her eggs are fertilized, she feels her brood growing too big to fit under her arms, and she rises to the surface, presses two of her arms against her body, and squeezes out tacky "toothpaste" made of white lime. As this flows out of her she blows it up like a big balloon, in the form of a fluted spiraling shell, and deposits her brood of baby argonauts in the center of the spiral.

The size and superb whorl of the argonaut shell resemble that of the chambered nautilus, so it is often mistaken for another species of that gorgeous octopus. It is equally prized by shell collectors, who call it *paper nautilus*. But this shell is not divided into compartments. Its texture is very thin, lightweight, and pure white, in contrast to the brown banded, heavy porcelain of the chambered nautilus. Until recently biologists, as well as shell collectors, assumed that this was the home of the argonaut octopus, all the more so because its aspect is that of a big, beautiful clam. Actually the argonaut shell is an exquisite egg-case, so light that Mother can hold it erect in the wind, where it acts as a sail taking the argonaut children on an ocean trip as fanciful as that of Edward Lear's owl and pussycat.

The maternal instinct of the common octopus is to cradle

and protect her babies in the armpit under one of her eight arms. Likewise, the argonaut does not abandon her offspring to the hazards of wind and waves; she holds on with her suction cups to the bobbing little sailboat and goes along on the trip with the children—she may even climb aboard.

THE GIANT OCTOPUS

In *Denizens of the Deep*,* Frank Bullen related a spectacle he witnessed from the deck of a whaler in the Malacca Straits off Singapore:

> There was a commotion in the sea where a large sperm whale was locked in deadly conflict with a squid almost as large as himself, whose interminable tentacles seemed to enlace the whole of his great body. . . . The head of the whale seemed a perfect network of writhing arms. . . . By the side of the black head of the whale appeared the head of the giant squid, as awful an object as one could imagine even in a fevered dream. . . . The eyes of the squid were very remarkable for their size and blackness which contrasted with the livid whiteness of the head.

Bullen at 18 had shipped on a whaler for adventure. He must have felt the romance of the Far East that moved writers of the time, like Kipling and Conrad. Everything about that fight to the death between a sperm whale and a monster squid-octopus smacks of vivid imagination—however, when he wrote about it years later, he was a biologist who specialized in whales. It seems that Bullen is one of the few people who have seen the giant squid as a wild animal in the ocean, and perhaps the only man who ever saw it in actual battle with a whale.

The giant squid became so intertwined with myths and legends, deeply rooted both in Scandinavian and Greek and Roman literature, that it lost reality. Linnaeus, the historic namer and systematizer of the natural world, shows the con-

---

* From Bernard Heuvelmans, *In the Wake of the Sea-Serpents.* Translated by Richard Garnett. (New York: Hill and Wang, 1968).

fusion about the giant squid by at first including it in his *Systema naturae* in 1735, and then leaving it out in later editions.

Through the nineteenth, into the twentieth century, occasional sightings of "the Giant octopus which the ocean brings forth at times from its depths to challenge science" came from reliable witnesses, some of whom swore in all sincerity they had seen it. But, lacking tangible evidence that could be measured and weighed and mounted in a museum, scientists tended to regard it as the "Thing."

The giant squid melodrama underscores my thesis that life on planet earth as created in the ocean is more fanciful than anything human imagination can conceive. We are prone to look at little living things in the ocean—and what could be more fairy tale-like?—but here is an animal for which the whole ocean is a niche, where it has been hiding, probably for a million years. Man can scarcely find it, and it met its match only when cachalots (sperm whales) evolved from land animals—even so, it is an even draw.

The scientists at last had their tangible evidence when they cut open the stomachs of cachalots and found enormous octopus arms. When the whole animal was reconstructed, the thing was 50 feet long—it still seemed to be a nightmare. Finally, around 1933, all sorts of giant arms, organs, and heads with 15-inch eyes were washed up on the rocky shores of Newfoundland. Nobody knows what happened—perhaps a turbidity avalanche down the continental slope sent a cataract of rocks and mud tumbling from the shelf, or perhaps a terrible earthquake brought disaster to colonies of giant squids cuddling in a canyon in the abyss. Thereafter the American Museum of Natural History listed the giant squid in its three-volume set of books, *The Animal Kingdom*—"total length over 50 feet . . . descends deeply and few live ones have ever been seen."

The regular little squids, 6 inches long, are well known and studied with wonder. They reproduce very well with hectocotyluses. Their eyes, like human eyes, can focus near and far. They have lobed brains enclosed in a skull made of cartilage (the creature pre-dates the fishes with true bones). Each of the fourteen lobes of the brain controls an emotion or muscle action. The nerve system has huge fibers which carry messages from brain to muscles fifty times faster than ordinary nerves because they are fifty times thicker than the nerves of other animals. The lightning nerve signals trigger the jet propulsion of the squid—which can shoot forward or backward as the fastest creature in the ocean.

The magnificent eye, nerve, and muscle system operates color spots (chromatophores) so fast that the animal can take on the color of its surroundings and disappear in one-third of a second. One instant it is silver gray, the next orange, crimson, or purple. It can even be black or white.

Squids which spend most of their time in dark depths are equipped with luminous spots all over their eyes and body. A squid may have five light organs on each eyeball. The middle organs of the eye shine with ultramarine blue and the outer parts with pearly sheen. "A squid is adorned with a diadem of brilliant gems!" exclaims a biologist. One squid that was left in a bucket of sea water on the deck of an oceanographic research ship was observed at night "sending up rockets for help"—it was expelling sparks "with phosphorescent splendor, making this squid-octopus one of the most brilliant productions of nature."

As this is being written, I receive a report in a scientific journal headed *octopuses struggle to control their prejudices.* They have curiosity. They make decisions.

I have placed the octopuses at the summit of organic evolution in the ocean, and pointed out how many of their characteristics, both physical and emotional, are distributed among land animals. When the era for life to emerge from the sea arrived, the pioneers upon the forbidding beaches would need, and take along, superior equipment like this.

But an octopus kind of thing couldn't breed organisms for emerging. It is irrevocably an ocean being. So key land animal characteristics were put to the test by crabs and snails caught between the tides, and by amphibians such as tadpole-frogs, and salamanders that had their elongated bodies and bilateral sym-

164

metry nerve and blood systems, eyes, and above all, male and female sex, from the bony, backbone fishes.

One other phylum, a generally unappreciated one, which emerged upon the land at some unknown epoch, is the *Annelida* (the segmented worms). Compared to reptiles, birds, and mammals, these look like a dead end. Their ocean origin seems fairly mysterious. What, if any, heritage do they have from the octopus?

There is no way to prove an answer as a scientific fact. The evidence dissolved away in sediments 600 million years ago. However, a fanciful hypothesis has been proposed by some marine biologists that the worm type of animal—still very numerous in the ocean—evolved from a hectocotylus! You will recall that this organ sometimes separates from its male to become "very restless and wind and twist in a determined manner."

A New York State College of Forestry report estimates that one acre of soil under a wild forest of eastern hardwoods holds three-quarters of a ton of earthworms. They thrive in the dark, damp soil and fertilize the soil by eating it and defecating. Tunneling worms also make soil porous, letting in rainwater. This makes plants thrive, and they in turn support more animals. If blind, stupid earthworms evolved from winding, twisting hectocotyluses, this was surely a fine gift to land life from the octopus.

14

Preparations for Going Ashore

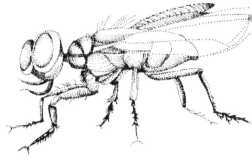

What superb melodramas nature stages! In her greatest masterpiece, "Escape from the Ocean," the theme is a mission impossible, the situations are fraught with suspense, the scenes are majestic with surf crashing against rocks and spreading on broad sandy beaches, and backdrop is a composition of sparkling, dripping, green seaweeds draped on rocks around low-tide pools.

Those seaweeds—everything had to wait for their overture—are an elaborate climax in the rise of the algae, through population explosions of diatoms and their kin in the Plant Kingdom, which supported the whole Animal Kingdom in the ocean with their photosynthesizing.

On the time scale of epochs we watch seaweeds roll in wind-blown wracks across beaches, green scums coat brackish estuaries, and fresh-water liverworts and mosses and ferns evolve —continuing to use traditional swimming sperms to fertilize their eggs. Plants, no less than animals, spur their evolution with sexual reproduction. When tree trunks raised green leaves into the air, the first true air-breathing, sunlight-loving land organism was created. With its own peculiar arm of the sea, it was ready to march inland.*

---

* Rutherford Platt, *The Great American Forest* (Englewood Cliffs, N.J.: Prentice-Hall, Inc., 1965), pp. 72–81.

166

Sap contains solutions of the same key elements of life—sodium, chlorine, phosphorus, magnesium, iron, etc.—which makes the ocean so chemically creative. The ocean gets these elements by dissolving them from the rocks which form the ocean basins, and by rivers disgorging them after rainwater has pounded and flowed on continental rocks. The sap streams of trees, the pioneers of inlands, get them from exactly the same source, the crust of the globe—with their roots in damp soil filled with sand crystals and water seeping up from the water table in contact with crustal rocks. Moreover, both sea water and soil water are fertilized by the bodies of hosts of microorganisms.

When plants adapted to all sorts of niches on land, threads of sap flowed through the whole diversity of trees, bushes, vines, wildflowers, and grasses. These carried both food factories and water over the land, and, in the course of epochs, to the most unlikely places—to Polar regions and mountaintops, in flat, crusty lichens and in inch-high flowers that can snuggle around stones and under snow for protection against bitter winds; to deserts, in plants with tough, leathery leaves which keep their sap from evaporating in scorching sunlight; to far-flung shores, with devices like coconuts that ride in ocean currents designed to be rolled across beaches by spreading waves.

Thus the descendants of the ancient algae unlocked the precincts of the continents, and the animals followed in their wake.

ON THE THRESHOLD

In the plankton world circling the globe in the primeval ocean all living things were similar in size and way of life. This was so because they shared the same environment in the ocean mixing bowl. After the plankton larvae perfected two-sex reproduction, bigger, stronger, more aggressive creatures turned up and many of them crowded into nourishing seaweed jungles along the shore. Then niches—places more or less insulated from the general ocean melee—played critical roles in creating creatures with exceptional aptitudes, whose descendants would use those aptitudes with their special equipment to spearhead the drive to the land. I visualize them as becoming desperate when their niches changed—under pressure to survive, they evolved snails and barnacles and vertebrate animals that would retain enough salty fluids to exist out of water.

Just before those between-the-tides adventurers, when no creature had yet arranged to live for a minute out of water, there was that crowd at the shore—and, astonishingly, one of their number has survived to this day to tell its tale.

By weird chance *Lingula* (little tongue) managed to live on and on in the same kind of niche for 425 million years, as though it was never pressured to evolve. We catch a glimpse of a thing that looks like a little fingernail attached to a rock underwater close to the shore of a strange planet. Its contemporaries vanished from the planet millions of years ago, leaving only their fossil imprints in shale. In Lingula's time there were no clams and oysters which today blow bubbles happily in mudflats at low tide, no snails or barnacles that cling to rocks waiting for the tide to come in. There were no crabs, no insects, there was no backbone or tooth, nor any moss or fern or any kind of land plant. All these came later.

So Lingula may be considered a live exhibit of what was shaping up before life emerged from the ocean. It has two shells, but it is not a bivalve, it is a dead end of evolution. The shells are horizontal, top and bottom. Those of oyster, clams, and mussels in the native mud or sand are held vertically on the right and left sides of their animals. Entirely different bodies had to be developed for excursions ashore. But note what Lingula does have, a few key organs which would help our ancestors cross the beach. It has a stomach, intestines, a little muscle called a "heart," that contracts suddenly to keep internal body fluids moving. Moreover, this humble, artless, shapeless thing has separate sexes—there are male and female Lingulas.

Millions of years after Lingula,* the wonderful backbone fish of the sea had stomachs, intestines, and organs for separate sexes

---

* Sequences of organic evolution among phyla were perceived later in fossil records. In the vague stretch of billennia on the threshold, the order in which land forms came is unknown and unimportant.

in their superb vertebrate bodies. And now the heart was a lobed, pulsing muscle that sent red blood circulating through intricate vessel systems. By this time we can suppose that snails —whose descendants would someday be earthworms—were gleaming in wet seaweeds; and that the dynamic Crustacea phyla —which would produce the insect hordes—were whitewashing wave-splashed rocks with barnacles.

Eras after Lingula established its immortal image, a backbone fish, but one unlike those of our time, was delved from dark depths, and showed a fish caught in the very act of turning into a land animal.

MISS LATIMER'S FISH

On December 22, 1938, Miss Courtney-Latimer, director of a small natural history museum on the east coast of South Africa, was in her office when a native fisherman brought her a hefty, 2½-foot fish. What was it? He was a veteran fisherman who realized that he had never hauled in a fish like this before.

Miss Latimer couldn't name it, but she became excited when she discovered features like those of fossil fish extinct for hundreds of millions of years. So she bought the odd fish and took it to her museum to ponder.

Referred to higher authority, the creature was indeed identified as a stranger that had lingered some 300 million years on a planet called earth where certain ocean creatures were shaping up to cross the threshold to the land. Honoring Miss Latimer for her perception that saved this thrilling discovery, the fish is named *Latimeria*.

If the evolution of Latimeria had not been arrested in a commodious niche deep in the thermocline, an environment virtually unchanged for half a billion years, or if the fisherman had taken it home to eat, science could only compare bony fish with land vertebrates and speculate on their relationship.

As it is, here is a purely ocean animal with the outward appearance of a regular fish, but instead of normal front and rear fins on its underside, it has two pairs of stumpy fleshy pegs that are actually legs developing. The animal also has a large swim bladder filled with fat—which is identified as the origin of the lungs of land-living, air-breathing creatures. Moreover, the nose is not like that of regular fish—it is a strange protuberance. Most interesting, there is a peculiar joint in the skull that gives it two movable parts. Shall we call this accommodation for a two-lobed brain?

This exciting discovery produced large cash rewards for more fishermen who might catch more Latimerias, and it sent scientists trawling the deep ocean off the coast of Africa. By 1953 the great Latimeria hideout was discovered, and more than twenty of the "half-baked" animals have been brought up. There is now no doubt that Latimeria is a living sideshow of doings at the threshold, when the new backbone fish in the primordial ocean were about to give rise to amphibians and thus—in the words of a report from the Peabody Museum, Yale University —give rise to all the tetrapods including man.

Latimeria's hideout, by the way, is a black, huge, slowly moving whirlpool, located 1,500 feet or deeper below the fast, extremely treacherous currents of Mozambique Channel between Africa and Madagascar.

HALF IN, HALF OUT

We still see today many half-outs still struggling to gain the land. They are a remarkable spectacle of creatures which contracted for the survival assets of both water and land.

One is shaped like a horse's hoof. I have watched many horseshoe crabs emerge out of Great South Bay to lay their eggs on Fire Island, New York. It staggers across the gray sand among the reeds like a small boy who has left his clothes at the top of the beach and is trying to sneak to them under an upturned basin. The animal has "gill books" by which it can live both underwater and on land, pincer claws like a lobster, compound eyes like an insect, four pairs of legs like a spider. The horseshoe crab lives almost all year underwater, plowing up the mud and sand for worms, shrimp, clams—until June, when it emerges to find a sun-warmed spot to hatch its eggs in sand at the edge of a meadow.

The female is in the lead. She may be 20 inches across, while the male is a 6-inch pygmy. She advances resolutely toward the top of the beach, dragging along her puny mate. He has holding hooks at the tips of his front legs that will leave scars on her tail. Since the performance takes place once a year, the age of a horseshoe crab can be found by counting the scars made by boyfriends holding her tail. The combination of security under water, and security from gulls *et al.* under their big shell basins, with the fast hatching of eggs by warm sun, has enabled this ancient pioneer upon the land to live through some 400 million years practically unchanged.

Similarly, turtles equipped with armor-plated domes have enjoyed the advantages of both land and water. The Galápagos tortoise, which weighs a quarter of a ton, and may gain 45 pounds a year, is an impressive creature, but it is a newcomer compared to the horseshoe crab. The tortoise dates from the Age of Reptiles a mere 200 million years ago.

It is solemn to see the struggle to cross the threshold still going on in our space age. There is the 10-foot Komodo dragon discovered in 1926 on a remote little island of Indonesia by Douglas Burden. "He looked black as ink, his bony armor was scarred and blistered, his eyes deep-set in their sockets looked out on the world from beneath overhanging brows. . . . A true dragon was there. We had found him. He was valid—an unforgettable picture of the primordial past of this extraordinary world."

There are the alligators, and the chameleons that change their body colors from leaf green to brilliant reds, to aquamarine, as though their genes had been imprinted with the memory of the spectral colors of the water where their ancestors were shaped. The water snake's undulations are those of waves and ripples. The water snake anticipates the method of mammals in a later age by producing live babies inside the body instead of inside eggs. Nevertheless, water snake uses the law of averages with multiple production. Forty-four little snakes may wiggle out of their mother in one night.

AND BACK AGAIN

Odd as it seems, bony fish are believed to have had their origin in fresh water in an epoch *after* crabs, snails, and amphibians pioneered the Great Emergence. That fish of inland waters should return to the spacious, luxurious ancestral home is not so singular. That is also what warm-blooded, air-breathing mammals like whales and dolphins did, when they wanted more security along with good eating and buoyant mating.

When you look seaward from the shore, you see the blue water stretching across the horizon with no hint that this area of the sea is separate from the vast ocean beyond. The continents have big shoulders called continental shelves, which vary in width. The shelf off our middle Atlantic coast is about 50 miles wide, it is perhaps about 15 miles wide off the Pacific coast, but it extends some 600 miles from the mainland on the Grand Banks of Newfoundland. At the seaward brim of the shelf, the continental mass plunges steeply vertical miles to the abyssal floor.

Geologists say that this dramatic feature of the earth's crust is the gift of the last Ice Age. At its peak, some 35,000 years ago, so much water was locked in continental glaciers that sea level was 433 feet lower than the land. The elastic uplift of land masses ensuing when they were relieved of the weight of ice left this shelf formation.

The point is that the continental shelves now under the ocean were shoreland once upon a time. In places, rivers crossing it cut fabulous canyons—one cut by the Hudson River has the dimensions of the Grand Canyon of the Colorado. Only 50 miles from the Verrazano Bridge of New York harbor is one of the most spectacular canyons on earth—which tourists will never see.

The ocean water upon continental shelves is much modified by the land. Sediments with vital minerals, eroded and dissolved from the land, pile up on continental shelves. Flora and fauna of estuaries, deltas, and river valleys enriched the shore waters with organic nourishment. The well-known marine fish of our menus are almost all caught where they throng in the comparatively shallow part of the ocean that lies upon the shoulders of the continents.

The bony fishes, evolved in fresh water, did not return to the ocean completely—as did some giant mammals. They were lured by the food in the shelf water, and by the opportunity to swim and spawn with more room than they had in estuaries and river mouths. What geologists see clearly happened in the last Ice Age must have happened in earlier Ice Ages—three earlier ones have been identified—when the ancestors of the bony fish forsook murky crowded inland water for the wonderful salt-water habitats where we find the finest ocean fish today.

The bony fish represent a stage of life midway between the ancient ocean plankton world and ourselves. Let us try to envision that world.

Soon after the between-the-tide creatures such as the horseshoe crabs had won a beachhead—led by some primitive mosses, ferns and liverworts put ashore by the seaweeds—some characters turned up that had extraordinary energy and capacities. They had inherited genes like those which gave sea squirt larvae notochords along their back, and other genes which gave octopuses their fine eyes with lenses and retinas, and others that bestowed blood vessels in squids, and gave lobsters muscular hearts, and male and female sex.

Such a combination made a strong, independent creature, that could buck outgoing tides and river currents. To that end some of the more up-and-coming among the primitive bony fish took the water route inland while they acquired, by the law of natural selection, torpedo-shaped bodies with sense organs at the front end and a central nervous system extending from head to tail along the back.

Man's body is like that of an ape, and the bodies of both man and ape *have the same basic design as the body of a dogfish.* Both have skulls which house brains, vertebrae backbones which house spinal cords, hearts, stomachs, livers, kidneys, spleens, ovaries, testes, glands, and hormone systems all in corresponding locations. A skeleton of a garfish, a fresh-water fish that is 5

173

feet or longer, was once dredged up from the mud in the Mississippi River and mistakenly identified as the skeleton of a boy who had drowned.

We can suppose there was a time when fish that reached farthest into estuaries and fresh water at river mouths found themselves in a tight spot when a drought dried a reedy area and reduced the volume of run-off water coming down a river. Then natural selection spurred the converting of swim bladders into lungs, made kidneys more efficient to eliminate body fluids and so regulate their osmotic pressures, and caused fins to be used for walking.

Such a performance is not so imaginary as it sounds. The African lungfish must come to the surface frequently to breathe air through its gills—or drown. Garfish come to the surface to gulp air when the water becomes too foul, or too warm. The famous climbing perch in rice paddies of the Philippines is often seen to quit its underwater home, shift to air breathing, stagger overland on its fins, and even occasionally climb a tree. Mud-skippers, among the stilts of mangroves in tropical swamps, are fish acting like frogs. They leave the water to hop around on mud flats. With the aid of pectoral fins and a muscular tail a mudskipper fish can leap a yard.

When the ancestors of the bony fish gave up the struggle to survive in fresh and brackish water they returned, not to the deep ocean, but primarily to its fertile borders on the shoulders of the continents where plankton mingles with organic bounty from the land. No wonder marine fish are beautiful and delicious.

Whales are an interesting contrast. Their ancestors were warm-blooded mammals that walked on the land, in an unknown era, and took off through the surf and headed for mid-ocean. In the big deep they became the biggest animals that ever lived on earth—a blue whale with a 150-ton body is three times bigger than the largest dinosaur. Whales became "stuck with" mid-ocean both for cruising range (the sea upon the continental shelves is only 8 percent of the global ocean) and for diving depth; a whale may go down 3,200 feet.

What about food out there? Fish and game animals—such as seal, dolphin, sea cow, sea lion, walrus, manatee, and dugong—

are in the shelf waters. So whales dispensed with fish and game. Instead of teeth they have sieves in the back of their huge mouths. A whale rushes through a mile of plankton, mouth open wide; while a torrent of soup pours down its throat, the sieve collects millions of plankton creatures. Its favorite food is called "krill," shrimp about an inch big. This is indeed nourishing fare—a whale calf suckled on its mother's milk grows at the rate of 200 pounds a day!

Whales, and also inshore seals and dolphins, are an archaeological mystery. One would suppose that their huge skeletons would have left a fossil somewhere in the ocean floor. But sediments are shifting, and bones dissolve in a matter of epochs. There is no clue to explain the astounding fact that certain mammals returned to the sea. It is a good guess they were flesh-eaters which invaded streams and coastal waters like minks and otters in pursuit of prey. When they discovered this was a fine way to have an abundant life they kept right on going until, in a million years or so, they became adapted to marine life.

Not quite so mysterious as why and when whale ancestors took to sea is the way the giants adapted mammal reproduction procedures to the unstable situation in ocean rollers. The late Roy Chapman Andrews of the American Museum of Natural History related a rare eye-witness account:

An amorous bull whale executed a series of acrobatic performances evidently with the object of impressing the female. He stood on his head with tail and fifteen feet of body out of water. The great flukes waved slowly at first and then faster until the water was pounded into spray and the terrific slaps on the surface could be heard a mile away. This performance ended, he slid up close to the female rolling her about and stroking her with a flipper. Then he dived. . . . He was gone for about four minutes, then with a terrific rush he burst from the water, throwing his entire body straight up into the air. Falling back in a cloud of spray, he rolled over and over with his mate while he clasped her with both flippers. Finally, both whales lay at the surface, blowing white vapor spouts, exhausted with emotion.

Thus was love in the ocean pursued by warm-blooded, air-breathing animals like ourselves.

INSECTS—THE ORIGINAL LANDOWNERS

When you switch on the light suddenly in the bathroom, you may see something exciting, and usually disdained, in the tub. It is a *Thysanura* (little tassle), called silverfish, though it is not a fish at all. It darts and runs on the vertical wall of the tub like a housefly, outwits your swats, and you hurl it down the drain in a rush of water from the faucet to be drowned, or carried away as a helpless mite in the torrent—but not so!

You are seeing the most primitive insect alive on earth. The thing has not changed much since the original insect chassis was perfected (before the winged models). It adapted to our world by taking possession of dark niches under wallpaper and in book bindings, where there is starchy paste on which to nourish the children—yet it never discarded ancestral affinity for water. That rush of water down the drain offers a good way to escape fast. It would soon find a clinging place in the black drainpipe, probably with a bit of organic matter to eat, and then, when it is again dark in the bathroom, it will pop out much refreshed by the rush of water and food in the drainpipe.

This is a nice episode in the relation of insects and water.

After those first venturers upon the beach and at the head of tidewater in estuaries, it took 50 million years for plants to build dynamic swamps with their decaying debris in depressions of sterile rock. This is the period which geologists call the Devonian, because the fossil records of the first land life on earth were exhumed from the heather-covered moors of Devonshire (known as the Lorna Doone country).

The crustacean type of creature (which includes barnacles, crabs, etc.) succeeded during that time in bringing forth the triumph of the first insects, which we encounter in the Coal Age. They were oversized brutes, dragonflies with 30-inch wingspreads, mayflies 5 inches across, cockroaches 4 inches long. Spawned in the black muck with them was a hideous ancestor of the salamanders called "roof head," sunk out of sight in the muck with only its nose in the air. Four-legged land animals had to wait about 50 million years after the Coal Age swamps before they took a firm foothold on the continents in the Age of Reptiles.

Meanwhile, those aerial, buzzing, twig-running insects of the Coal Age are entitled to be called the first distinctive kind of land animal life. Insects have proved to be the most successful size and style of life, able to reproduce with population explosions while never confronting food shortages. They would be justified in calling us mammals newcomers and trespassers with only a presumptuous title to the land.

The descendants of those husky pioneer insects which imprinted the coal measures, to this day, 270 million years later, still return to breeding places in water or muck, as we noted in Chapter 10.

The dragonfly, queen of the air, has exquisite slender wings flashing the iridescent colors of sunshine. When egg-laying time comes, she feels the irresistible magnetism of a pond, zooms down to touch the tip of her abdomen to the water's surface and discharge little volleys of her eggs. Or she may alight on a water reed, crawl down the stem and laboriously insert her eggs with a kind of hypodermic needle into the stem below the surface.

When the eggs hatch, the dragonfly nymph (its immature form) emerges as an ugly, dark brown one-inch monster. It walks clumsily in the muck on the bottom and holds motionless; invisible to prey, it seizes them with a lightning stroke of its lower lip, which is shaped like vicious pincers. This terror of the pond, offspring of the most beautiful and graceful of insects, moves through the water with jet propulsion, sucking water into its abdomen and shooting it out like a bellows. It has gills and breathes like a fish.

Mayflies are better known to fly fishermen than to other people, for their drakes and spinners are usually designed after these aery insects whose slender, curving bodies are so attractive to fish. Mayflies are prime evidence that the first kind of life to inhabit the land were insects which evolved as sheer reproduction mechanisms in the air and sunlight. Mayflies have no mouths for eating, no organs for digesting. They take no food in the single day which is their life span.

Just before sunset you may catch a glint of their mating dance as they reflect the horizontal rays of the sun like copper sparks. A mayfly mounts straight up about 30 inches, then using its wings as a parachute it drops straight down slowly on the same line. Then up it goes again. The air is filled with countless numbers and frequently the vertical path of one may have its lowest point about halfway on the path of a neighbor. When mayflies find themselves in this position they synchronize their wings, so that both are on the same beat, going up and down together. Thus, males and females cling together and mate as they dance.

One crepuscular dance is usually enough to serve the purpose but some mayflies live to a very old age, on their time scale, to jig at a second sundown. After a successful dance the male dies and blows away like chaff, and the female hovers over water like a helicopter and drops two packets, which break up while sinking, so the released eggs reach the bottom separately, and quickly hatch soft crawlers on six legs. These creep around for several years until they rain upward through the surface, unfurl their pretty wings at the touch of air and catch the rhythm of the dance.

If you judge the success of a species by its numbers—well, let us look at western Lake Erie, where the depth and temperature of water, and the day-night rhythm, is perfect for mayflies.* Edwin Way Teale in *Journey into Summer* wrote a classic about a "mayfly storm" at Sandusky, Ohio. During its height, "truckload after truckload of the frail insects were hauled away from the city streets" while merchants turned off their neon lights (which lured and excited the mayflies like the red rays in the setting sun); outside, the housepainting was suspended because mayflies made the house furry, and traffic was snarled by slippery, crushed bodies of the insects.

The other of those air-breathing licensees of the Coal Age, the cockroach, has adapted to a simile of its black swamp muck

---

* Rather, it was a few years ago, before most of Lake Erie was killed by pollution.

for breeding in our nice warm homes, namely, the dirt from food that gets into crannies. It must be dark, nutritious for raising little cockroaches, and above all *moist*. Evans says that a cockroach is "a marvelous beast" * and opines that the American species came over in the *Mayflower*. He finds this creature, which is so beautifully adapted and adaptable through hundreds of millions of years, more interesting and exciting than news about people today.

The cockroach is fast because it has giant nerve fibers that send impulses direct to its legs, instead of roundabout via a brain. It is strong, fends insecticides, is hard to kill by dealing blows. Even when decapitated it runs, and it will live for several days until it dies of starvation. Yet note that it must be near water for breeding. The modicum of moisture retained by the "muck" in a dark crack is enough.

The modern housefly lays its eggs in warm wetness of manure, rotting vegetables, or kitchen slops. It has recently been found that their sustenance is chiefly the bacterial populations in those places—and bacteria are always underwater organisms. The buzzing, acrobatic flier in your room is adapted to live without any water visible to your eyes, simply by taking it in through twenty *spiracles* (open pores in its armor) from water vapor in the air; the rest of the body is covered with wax to prevent evaporation.

Insects have multiplex resources for keeping in close contact with water—although many, as in the case of the housefly, escape detection. A leafhopper seems utterly gay and free in sunshine and air, as it flies and leaps in the grass or on tree leaves. Transfixed on a leaf, it forms its two wings into a roof that sheds rain (and insecticide sprays). It does not need raindrops because this insect house has plumbing to tap the leaf's sap reservoir. Under the roof it drinks its fill with a sucking beak, and it lays eggs which have tiny projections on their undersides that pierce the juicy leaf like a faucet to supply leafhopper embryos with good drinking.

So insects, of the phylum of the barnacle-crab-shrimp, never

---

* Howard E. Evans, *Life on a Little Known Planet* (New York: E. P. Dutton & Co., Inc., 1968).

179

abandoned the ancient ancestral water home, which offered all sorts of impulses for the evolution of species, and superior cradles for laying eggs and raising their young.* Today we see hot, humid swamps, and rank, wet grass or bushes, and pools of stagnant water, producing *clouds* of insects.

The reciprocity between the Plant Kingdom and earth's most successful land animals presents a highly imaginative jubilee of water and life. On the time scale of epochs this dazzling development is modern, because all the flowering plants, a botanical category that includes broadleaf trees, bushes, and grasses, evolved in the *Cenozoic*, which means "modern life era."

Magnolia and birch were among the first flowering plants on the globe—paleontologists call them "very ancient," yet they appeared about a hundred million years *after* the Age of Reptiles. Large, showy magnolias were plentiful from Martha's Vineyard to Texas a mere 50 million years ago. From then on all the colors and fragrance of flowers and the colors and liveliness of insects of our world arose together out of rivers of sap.

Scientists speak of this with objectivity, using phrases such as "insect pollination is the principal guide force in the design of flowers." But we can say that the Plant Kingdom and the Animal Kingdom have created an astonishing symphony of landlife by making arrangements for foliage reservoirs of water in all sorts of corners of the land. Birds, conjured up by feasts of delicious insects, add music to the show!

Of course, the success of some insect forms of land animal is a bit disconcerting. Just see them—leafhoppers, thrips, aphids (an entymologist estimated there were 21 million aphids in the canopy of one cherry tree), plant lice, caterpillars, and spittle bugs which suck up water that inflates the palisade cells of a green leaf, stirs the water with body fluid, then blows the "soapy" mixture out its anus, to make a frothy, humid green-

---

* The underwater lives of the commonest land insects, such as mosquitos, dragonflies, and mayflies, is as exciting as fiction, and would fill this book. Top entomologists have had a field day with this subject. For example: Frank E. Lutz, *A Lot of Insects* (New York: G. P. Putnam's Sons, 1941). A. D. Imms, *Insect Natural History* (London: Collins, 1947). Malcolm Burr, *The Insect Legion* (London: Nisbet, 1954).

house in which spittle bug children can grow up in luxury and security.

Many true-life "fairy tales" of flowers and insects have been told in solemn scientific lingo as in poetry. One of them concerns a particular wasp and an orchid. The orchid has evolved its color markings, and the size and arrangements of its reproducing mechanism (stamens and pistils) to deceive a wasp, which recognizes the flower as a female of its kind. The wasp enters and copulates with the orchid, thereby pollinating the species.

The most famous case is, of course, the bees and flowers. And those who cultivate figs for the market well know that only one kind of insect can pollinate the edible figs—its magnificent name is *Blastophage psenes*, which translates as: "Capri fig insect eat blossom."

The gravamen is that the organs of flowers are *the most reliably moist spots in every neighborhood.* They are also the right size to fit insects. Insects have plastic bodies which can be quickly shaped in fast generations to fit any shape of niche. They have prospered with flowers—especially the swollen ovary at the bottom of the pistil, which is filled with lush sap—for the same reason that they prosper in ponds, swamps, and foliage.

Let me underscore this with the strange tale of the yucca moth. Springtime visitors to the desert of southern California see the mountains abloom with towers of white flowers 10 feet tall—from a distance they look like giant candles placed to decorate dry, sandy hillsides. This landscaping of desert real estate is entirely the work of *Tegeticula alba*, meaning "little white mat." * The yucca, often called "the desert candle," is a botanical impossibility. The plants stand solitary, maybe hundreds of yards apart. Their pollen is heavy and sticky; it cannot blow in the wind or catch in the furry bodies of insect visitors. The plants, with big stiff swords for leaves, stand immobile, unchanging almost the year round, until suddenly—beginning on the

---

* A species name is often based on some feature that strikes the first biologist who describes it. C. V. Riley, Missouri Botanical Garden, gave the yucca moth this name doubtless because he first saw its white wings plastering the ovary of the desert plant. It gives no hint of the weird love story.

same night every spring—a tall, straight stalk mounts swiftly to a height of 10 feet or so, and blooms with white flowers. A day later the flowers have wilted and fallen off, and the great stalk that bore them crumples.

On the very same night in which the yucca bursts into bloom the desert air is alive with little fluttering ghosts. They are the yucca moths that have emerged from under desert pebbles; all are females bearing loads of fertilized eggs. Each makes straight for a flower, drills a hole in its ovary, which contains unripe seeds, and pushes in her eggs, wadding them carefully into the moist womb of a yucca flower.

The pistils of the same flowers are not yet ripe with pollen. This is a provision of nature that insures that yucca flowers will not be self-pollinated. So, to make sure that the flower which holds her eggs will ripen its seeds and nourish her babies when they are born, the yucca moth flies away over the desert in the cool night air to another desert candle calling to her in the dark of night with its white flowers. She seeks until she finds flowers with *ripe* stamens, and scrapes off the fresh, sticky pollen. Patting this into a ball, as a boy makes a snowball, she heads straight back to the tiny target, perhaps a quarter of a mile away, where she deposited her eggs. Climbing to the top of the same yucca, she rams the pollen into a little hollow on the pistil, just the right size to receive it. When her eggs hatch, the larvae will find themselves well provided with pregnant, moist seeds to grow on—and the baby moths always leave plenty of seeds uneaten, an inborn trait that insures the survival of yuccas for moth descendants.

In so many ways insects in our day have achieved preeminence as the landowners of planet earth—whether in the way of a dragonfly or housefly, in the way of a cockroach or mayfly, or a yucca moth. And never for a moment are they out of touch with water, even in a desert!

Rivers never flow in a straight line. (*T. G. Freeman, Courtesy U.S.D.A.*)

Snow surveys carried out by the U.S. Department of Agriculture help determine how much water will be available for the coming summer. (*Courtesy Soil Conservation Service, U.S.D.A.*)

Mountain snowfields are weather machines. They turn humid air from the valley into clouds. (*Robert J. DeWitz, Courtesy U.S.D.A.*)

The "burning tree," an aerial view of the Colorado River Delta, bears an amazing resemblance to a network of blood vessels. (*Aero Service Corporation, Division of Litton Industries*)

The most dramatic lake of our country. A mountain three miles high collapsed, leaving a huge fiery caldron. In the course of a few centuries spring water filled this to a depth of 2,000 feet. Crater Lake, Oregon, is gorgeous indigo blue.

This photograph of the face of the ice cap in Greenland illustrates the glacial delta of northeast United States in the making. The march of plant life is close on the heels of the retreating Big Ice.

Water dissolves granite, releases minerals, lichens take hold, flowering plants spring from the sediments.

No obstacle to a salmon on the way to its breeding place. (*Richard F. Shuman, Courtesy Bureau Sport Fisheries and Wildlife*)

15

The Call of the Land

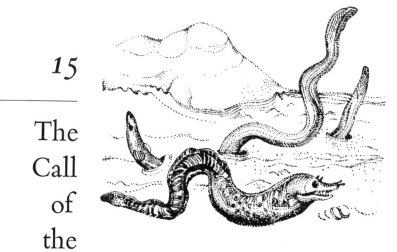

The big sockeye looked up at the waterfall that loomed ahead. It stopped about 8 feet from the crashing at the base, made a few turns to maneuver into position, flexed its hefty body, and with lashing tail in a frenzy of spray dashed straight toward the thundering obstruction in its path. At the touch of seething water the salmon's arched back emerged for an instant as it bent almost double—and then let go like a taut spring, sending the fish flying through the air, describing a beautiful hyperbola.

The running start, like that of a pole-vaulter, lent the momentum needed to clear the crest of the waterfall—as it looked to the fish. But the summit of a natural waterfall in a salmon stream is always higher and beyond the vertical drop, so it was no surprise to the full big sockeye to splash down in rushing waters pouring over ledges, and winding between boulders.

Instantly the fish turned on swim-power with throttle wide open. Vibrant and sinuous like the current, it matched its strength against the racing water; outweighing the relentless opposing push, it slid between boulders, and even swam upward at a steep angle to clear a ledge. With pauses to rest in eddies and in pools in the lee of boulders, the fish soon gained smooth-flowing water and resumed a leisurely pace upstream punctuated by frequent spurts to catch a minnow, circling to smell the water and get its bearing, and an occasional leap-and-splash to test its muscle tone.

A salmon headed inland may reach its goal in the lower reaches of the river, or it may pursue its trek for weeks, fighting its way on and on upstream, the full length of a giant river system such as that of the 2,000-mile Yukon in Alaska, the 1,200-mile Columbia that traverses the state of Washington, and the 850-mile Fraser River of British Columbia.

The sockeye, which we just saw in our mind's eye make a fine leap (they can clear 8 or 9 feet over the vertically falling water) and then breast the tortuous rapids among boulders, was on its way to a particular branch of the river, and then up that to a particular brook, and then to its personal pond. It is going straight to the sediment in that pond where its mother spawned and it was born four or five years before.

Somehow the fascination of that spot is never lost, no matter where the fish travels across ocean horizons—there is a documented case of a tagged salmon returning home to North America after a 3,700-mile journey in the ocean. The irresistible urge to return to its birthplace comes when the salmon has grown up after living for four or five years as an ocean fish.

Consider the situation. Salmon grown big and strong while enjoying boundless freedom and all sorts of good food in the open sea, are now ripe for reproducing. If it is a female, it is heavy with ripe eggs. If it is a male, it is loaded with sperm. But, instead of proceeding to spawn then and there—or even finding shallow water near the shore for the purpose—they must swim the width of the Pacific, if necessary, and then fight their way far inland, leaping over waterfalls and fighting rapids, breasting the current every inch of the way.

Wherever Pacific salmon are solitary and scattered, far and wide in the ocean off the coasts of Oregon, Washington, British Columbia, and Alaska to the tip of the Aleutian Islands, to Japan, to Kamchatka, each one when prompted by its gonads faces in a northeasterly direction to join a congregation either in Bristol Bay (in the Bering Sea just north of the Alaska Peninsula) or in the Gulf of Alaska off Kodiak Island. Five different species of salmon may mingle in an assembly area, which serves as a freight-car sorting yard. They now sort themselves out, not just according to species, but also according to geography, steering a course to their native river systems. It is thought that male and female sockeyes find each other in this stretch and that they arrive at their personal river mouths as couples.

This happens in early spring, when experienced fish watchers find salmon of splendid size in the ocean a short distance off the mouth of a river, where they seem to be swimming around aimlessly. They have arrived in less salty water, and they must now proceed into the river with great caution because their body fluids have the saltiness of the open ocean. A salmon may take a week or so while it makes its way slowly through the brackish water converting from a marine fish to a fresh-water fish. If it suddenly plunged into the river it would be drowned by fresh water flooding its organs that are in salty balance with the ocean. (Body fluids of a fish must be *isotonic* with the water in which it swims.) This heightens the wonder of the migration of salmon.

The beautiful salmon have long been celebrated in poetry and art, and in the hearts of their worshippers. The spectacle of a fish in orbit over a waterfall, reflecting the colors of the rainbow, is front-cover illustration. The cult of anglers founded by Isaak Walton in 1640 flourishes today with a jauntiness and lingo all its own. An angler can call a salmon by a splash—it is a *chum, pink, sockeye, silver, coho, chinook,* or *king.* If it is a young fellow that has not yet tasted the ocean it is a *fingerling* or *fry.* They tell you that "the king salmon is a savage fighter with a stout heart and tremendous fighting ability." Flytying is an elite art, and a flytier must be a keen student of salmon or trout. A top flytier tests his creations by lying prone, face up, on the bottom of a stream while holding his breath to get a fish-eye view of the design and sprightliness of a fly he has tied.

Yet through the long history of angling, with all the expertise, salmon have kept some profound secrets. One can see that they never spawn in the ocean or along the shore like almost all other marine creatures. Their breeding spots far inland in corners of ponds and in eddies of brooks have long been known precisely. The female deposits her eggs on clean loose gravel under running water. The male with a flourish douses them with sperm. The running water buries the fertilized eggs under inches of gravel. A month later the pond teems with fingerlings.

An angler knows where to take his position in March and April to see a big fellow come leaping and fighting fast water to reach the place where it was born. That place may be one of a hundred thousand lakes which gleam like eyes scattered among dark crowberry bushes of the tundra in Yukon Territory. To attain the spot, a fish must fight the push of water all the way across Alaska and find a certain cold, clear brook among lingering snow patches. Or the fish's birthplace may be more than a mile above sea level in the Rocky Mountains of British Columbia where threads of the Columbia or Fraser rivers tap snowfields and glaciers.

Fishermen curious as to how high a salmon climbs caught some in lower stretches, tagged their dorsal fins with aluminum markers, and released them. One was recovered 6,648 feet up in the Rocky Mountains which it reached after five days.

The rest of the salmon story seemed unanswerable. Who can watch a fish under the ocean waves? Who can follow the migrations of individual fish among millions? (It is a long way between Kamchatka and Yukon Territory.) What detector can tell how the brain of a fish works when it is selecting the right turn at a fork in the river?

Dr. Arthur D. Hasler, a zoologist at the University of Wisconsin, struck by the awe and wonder of salmon, exclaimed: "No one who has seen a chinook of 45 kilograms (99 pounds) fling itself into the air again and again until it is exhausted in a vain effort to surmount a waterfall, can fail to marvel at the strength of instinct that draws the salmon upriver to the stream where it was born."

Marveling thus, and thrilled, Dr. Hasler accepted the challenge of the salmon, which lead two lives—far inland among mountains, hundreds of miles from the coast, and beyond many far horizons in the open ocean.

He began with a ten-year study of white bass in the Wisconsin lake region. Bass are *Teleostei* (true bony fish like salmon). They are active and exploratory, and their ways are much influenced by past experience. White bass are localized in lakes, yet they migrate a few miles, poking up streams and idling among islands, but always returning to a personal spot in the lake or its tributary streams to spawn each spring.

Dr. Hasler and his enthusiastic research students went after the bass to get to know them as individuals. They plotted the spots where they were caught, tagged them with tiny aluminum disks stamped with numbers, recorded the date, hour, and location. They attached angler bobs to dorsal fins with fishhooks, and followed these in canoes, paddling as quietly as possible so as not to distract the fish, while charting the direction, speed, and maneuvers of the bass. Realizing that a bob may act as a brake, they substituted Ping-Pong balls, and these were white and easy to follow as they bounced around over the lake.

Recently a transistor was designed which broadcasts radio signals for a mile or so. It was small enough to be inserted in the chest of a fish without disturbance, and now the researchers could use an outboard motorboat to pick up the free movements of fish from a distance of half a mile or more. With transistors in the fishes' chests, the courses they took could be plotted day and night, in fair weather and foul. The charts revealed that the fish were guided by the sun by day—and by the stars at night!

The fame of Professor Hasler's fishing spread. Graduate students joined the chase. In 1966 he took his researchers to the Northwest, well-prepared to tackle the great enigma of the salmon.

THE SALMON TELL SECRETS

WHY are salmon hell-bent to beget far inland up in mountain valleys when practically all other marine life from lobsters to ocean fish easily spawn down on the coast?

HOW does a salmon remember a particular place where it was born in a continental wilderness—which it quit five years before when it was a small juvenile smolt—and how can it make all the right turns, avoid all the wrong ones, to get to that spot?

WHAT compass or chart does a salmon use to navigate in

the open ocean, despite wind-driven waves and ocean currents which may not be going their way, to arrive at a native river?

We are prone to think of memory as residing in the brain because to us memory is a conscious recall. But a salmon has an *odor memory*, which is keen, incredibly discriminating, and not dulled by time because it is imprinted in its cells. This is instinctive memory. We have odor memory, and with us it evokes mental pictures. Whenever I sniff the fragrance of the sea from afar, my heart leaps up and my mind's eye sees rocks draped with seaweeds around a low-tide pool where I played as a 6-year-old at Kennebunkport, Maine.

A salmon can have no mental picture of the place where it was born. It cannot remember in that manner because a fish has no cerebral cortex. But its nerve cells remember the smell of a place —and this evokes, not a mental photograph of a corner of a pond, but the "strength of the instinct" that Dr. Hasler marveled at. The odor that reached the salmon was carried in the river current. It came into the river in a tributary, into the tributary from one of its forks, into the fork from a lake, and it entered the lake from a nook behind a boulder, or from a pool at the mouth of a brook, where the mother of the salmon spawned in sediments.

Obviously the odor borne downstream could not go more than a short recoil current above a fork. The currents must carry the odor along a single path from its source. A salmon must take the right turn or lose the scent. In following this train up from the ocean the salmon is, in a sense, climbing the tree of the river. A river system on a map, as you know, is designed like a tree, with a trunk and many sinuous branches and twigs. In spring months the huge tree may be alive with salmon throughout—struggling through the network in all directions, each one with its nose thrust forward to smell its birthplace. The olfactory nerves of a salmon are concentrated in a pocket back of its nose. It always heads straight into a current, it never moves sidewise or backs up. When searching for a scent which it has lost temporarily, it describes circles.

Dr. Hasler likens this behavior of salmon climbing the river tree to that of a bloodhound following a scent trail as he backtracks, zigzags, searches, until picking up the scent again, he bounds ahead in the right direction. When the fish overshoots the mark and misses the scent at a fork of the river, it lingers in water close to the bank where the scent is less diluted than in midstream, then it retraces, crisscrosses, sniffs along both banks until it finds the direction from which the currents that bear the scent are coming.

The "inextricable" mystery of fish taking the right turns was solved by an experiment which needed no microscopes. The researchers caught a hundred salmon just above the fork of an important tributary. They were all tagged for identification and the olfactory organs in the noses of fifty of them were removed. All were then transported below the fork and released. Those with clear noses unfailingly made the right turn, those without smell organs were utterly frustrated, and took the wrong turn as often as the right.

This is a fascinating finding for an important scientific report —the very idea of fish smelling the right turn at each fork in the road! But it surely poses a curious question. What is the physical reality of the scent trail which the fish follows so unerringly? It is no answer to say that a fish has a much keener olfactory sense than people. So have dogs and moths.

We mentioned in Chapter 9 the odor particle with the new scientific name, pheromone, and the way it exerts its powers by being dissolved, as in the mucus of our noses. We noted how this was a time-honored invention for locating and summoning mates, with equal felicity in water and air. The case of Fabre's emperor moths showed the incredible power and precision of this invisible molecule.

But calling mates in the neighborhood with a speck of chemical substance, at a time when both are highly receptive, seems to have validity, whereas the salmon picking the right path through a labyrinth of unstable water currents, after years at sea, is a different matter. It is inconceivable that a dainty odor particle can keep its chemical integrity after years of dilution in hundreds of miles of a river system, to be greeted by a fish arriving at the mouth of the river.

However, this is what happens. It happens in the wide deep

flow of the lower reaches of a big river, it happens also in the rolling brackish water off the river mouth, and this faint and delicate signal even reaches across the horizon. To imagine such a thing had to await electronic science. This contemplation of the parts played by chemical solutions of water is an exciting frontier of today's biology.

Let us try to make scientific sense of this homecoming of salmon. Using successive teams of students on this scent trail project, with singleness of purpose and perseverance, Dr. Hasler found some answers. What gives a particular odor to a location such as a birthplace nook of a salmon? How does this "fingerprinting" of the spot last at least three or four years?

The salmon detect their personal odor molecules even as they are mingled in myriads of various odor particles, because the scent trail of a salmon carries the smell of the infusion of organic matter where it was reared. This odor molecule is made by dissolved roots, decaying bits of trees and bushes that drop in the water, stems of grasses that line the bank, the larvae and dead bodies of the insects at that place, plus dissolved mineral crystals from sand and pebbles and the kind of rock crystals that compose a boulder washed by the water. Can you think of anything more precious and personal than the odor of the water where a salmon is born!

This odor molecule is as delicate and intangible as the sparkle in a drop of water, but it is not evanescent. A pheromone holds its singular chemical virtue an indefinite time. It travels the length of the river, it spreads across horizons in sea water—the odor power is incorruptible. The migrations of salmon and all other fish and creatures of the sea—and of many ocean birds such as the Tristan great shearwater and the Arctic tern—depends on the amazing immutability of the chemical particles that impart both odor and color to water.

How does a salmon remember the odor of its birthplace? The ingenuity of researchers who experimented with infusions of plants and minerals from hundreds of salmon breeding places and discovered that each locality has its special odor, led straight to this related question of how does the fish remember that particular odor for years.

Inquiry into this subject started at spawning spots—in a hideout behind an island in a lake, behind a boulder, in a quiet spot of water under the bank of a stream, in pebbles of a clear flowing brook. The research students watched while the female spawned her eggs and the male spread his milt over them, and they became buried by sediments—and then in a few weeks the young fry popped out and the water teemed with them.

If the fish are netted as soon as born and transported to another stream, they are imprinted with the odor of the spot to which they are transferred—not that of the place of birth. In short, the memory of an odor is not genetic. They did not get it from their parents. It is imprinted *by what they eat* after birth when they are little fry.

Salmon can be manipulated to swim up a different river system. In one experiment 1½-inch fry were carried by airplane to a distant river—imprinting takes about a week's exposure to another site—and they fought their way to that place after returning from the ocean.

This is not brain memory. It is molecular memory, an instinctive memory that resides in all the cells of the body, instilled by what the baby fish ate in its first weeks after being born. The wonderful memory is exerted by protein enzymes inside cells, and hormones that carry the "memory" in watery body fluids to activate organs and muscles.*

"ON COURSE!"—IN THE OPEN SEA

What means does a salmon use to navigate to its river mouth far across many horizons? This sounds like an impossible question to answer. There are direct ways of solving the puzzle of how they find their birthplace after arriving in a river system. It is possible to catch fish at different stages and try, and try on for odor, "fourteen different plant rinses." But where do the salmon go in the ocean and who can see them heading home?

---

* Words like "pheromone" or "hormone" are convenient tags for utter mystery. Even trees and flowers have "hormone memory"—with *phytohormones* or *auxins*. These control leaf dropping, bud opening, root elongating, and sexual reproduction. Evidently a brain is not needed for this kind of memory.

The first step was to chart the areas of the ocean where salmon spend their years growing up, and from which they depart to resort to the continent. The interest of the fishing industry, plus an appealing problem of biology, led to the charter of four vessels which did some extraordinary fishing across 500,000 square miles of North Pacific Ocean from April to late July—the time of the year when the big salmon ready to spawn depart from their ocean assembly areas and head for their personal rivers.

In two seasons (1962 and 1963) each vessel made seven cruises, using a gear known as a *skate*. This is a nylon fishing line a mile long buoyed by floats at short intervals that hold it in a horizontal position as it snakes over the ocean surface while slowly being towed. Watching the floats from the vessel, the helmsman, with a biology professor at his elbow, can keep that line curving and reaching out maybe a mile, with 1,000 hooks dangling from it about 4 feet below the surface, each one baited with a little smelt.

The skate, with its mile of nylon and its thousand hooks, was hauled aboard at intervals whenever a good percentage of the floats observed through binoculars bobbed as though they had caught fish. Then the researchers left their coffee cups to tabulate the number of fish per thousand hooks, the size and kind of salmon, the latitude and longitude, then attach an aluminum identification tag, and set the fish free in the ocean. A reception committee was awaiting their arrival at the mouths of the chief salmon streams—the time intervals would show how fast they swim in the ocean.

The success of the skate survey in its first two years was a surprise. A good percentage of those tagged far away at sea were recaptured in rivers, brooks, and ponds. Exciting statistics began to reveal answers to the historic salmon enigma. The long-line skates had followed and gathered salmon from our Northwest Coast west to the Aleutians, and westward along that island chain for a thousand miles to offshore Japan—and onward across the Sea of Japan to ocean off Korea and Kamchatka, with Japanese cooperation.

The general picture is that all kinds of salmon from many far separated river systems assemble in definite areas according to ancient customs. Many individualists pass up the assemblies in the Gulf of Alaska and Bristol Bay, or swim away after a short stay, and keep going westward to join salmon in the Orient. But wherever they may mature scattered across the vast North Pacific Ocean, each will return to home base to reproduce.

Of course, the research teams on the four vessels were not merely servicing their skates, and getting quarry tagged and back into the water on their way to the scientists waiting along rivers and at the favorite brooks and ponds of anglers. They were keen about temperatures of the water at the depths used by the long-distance travelers, and the winds, and ocean currents, and especially cloudiness—can salmon migrate by taking a bearing on the angles of the sun?

The amazing answer is that they do! Every mariner knows that "shooting the sun" means observing its height above the horizon and the angle it makes with the horizontal azimuth at a given hour (marine navigating charts are based on noon). But these components that determine a position at sea vary every day as the sun gets higher in the sky in the northern hemisphere in the spring.

If this sounds complicated, to you and me—it is not to the salmon. One of the milestones in the history of biology is the discovery that migrating ocean salmon (and long-distance, over-ocean bird migrators as well) have a biological clock in their bodies. This is an inherited sense, installed in their DNA, by which it tells the daily sun time in accord with the season, and by which it perceives the altitude of the sun in order to know its latitude. Let currents and winds deflect a salmon from its course in mid-ocean—it corrects the deviations.

In one particular the salmon differs from men in taking a bearing. Instead of at noon, it notes its position at *sunrise*. Actually this is much more accurate—man's noon is put in broad time zones, and he uses arbitrary time like "daylight saving." A fish is not so unprecise.

Biological clocks in salmon are too complex to discuss in this volume. So let me authenticate them with a quotation from a report * by Dr. Hasler:

---

* Arthur D. Hasler, *Underwater Guideposts* (Madison: University of Wisconsin Press, 1966).

"A salmon can 'calculate' the bearing that must be taken to the sun, much as a human navigator employs his sextant. That an animal is actually able innately to accomplish something for which men require instruments, charts, and tables is quite incredible, but through experimental observations we know that this does, indeed, happen, and moreover, this ability has provided some of the strongest evidence of the biological clock."

WHY DO SALMON DO IT?

To say that salmon leap waterfalls, fight rapids, climb tremendous river systems to reproduce is no answer to *why*. The fish do not have a conscious purpose like a gypsy girl who drops behind a caravan to have a baby behind a bush. Consider the futility of the salmon way of life. The spawning spot may be a thousand miles inland in a tundra pond that holds a solution of roots and twigs, of balsam fir, larch, and wind-twisted black spruce at the Arctic tree line. And then the offspring, after lingering there for a year, go swirling downstream and fling themselves over waterfalls retracing the route which their parents laboriously took up from the ocean.

All the elegant techniques of science cannot find the why of anything in life. They can detect chemical structures of molecules such as DNA and the way they act, but this does not tell why they are as they are. However, wonderful *laws of life* have been discovered, and organisms seem to make sense when we see them conforming to these laws.

The law that interests us here is the traditional one perceived by Charles Darwin: species were the result of survival by natural selection. In other words, salmon must have evolved from ancestors who survived by dint of overcoming the great physical barriers to breed far upstream.

But why did their ancestors have to take such energetic steps to survive? Why did they have to breed *upstream?* The answer is that they didn't have to. They could have turned the other way and vanished in anonymity among myriads of ocean fish, while the few who tackled opposing forces in trying to get farther inland perished. In that case we would not see this unique and beautiful sideshow, this fantastic, superfluous product of the operation of the law of survival by natural selection.

Let us speculate on how this might have come about.

As we noted earlier, according to reliable evidence of geology and comparative anatomy, the bony fish originated in fresh water, not in the ocean. Their fine skeletons and organs were probably created by a speeding up of the evolution of their class in the face of difficulties which they met in fresh water.

Those which found themselves in really tough trouble had to cope with treacherous currents in the mouths of rivers, in tide rips, downrushing flood waters, shifting channels, seasonal changes of temperatures and volume of their home water. Those in catch-rain bogs and lakes fed by streams were presented with awful challenges to survive in a drought. Such conditions incited ancestral fish to jump and splash and adapt to changes of salinity.

In the age when salmon ancestors evolved in fresh water hundreds of millions of years ago, land masses were not as they are on the globe today. Continents have drifted apart, mountain ranges raised up and flattened by erosion, oceans invaded continents, and never-ending volcanic action rent the granite bone of the land. The Snake River, one of the fabulous salmon streams of our Northwest, cuts its famous mile-deep canyon through the Columbia Plateau, which is a block of volcanic rock of 35,000 cubic miles. Another revelation of the way continents have drifted apart and altered is the arc of the carbon swamp forest in the Coal Age, which shows that North America, Greenland, and Europe were a continuous land mass with a warm humid climate 100 million years *after* the first bony fish appeared on earth.*

---

* On three polar expeditions, I saw where the Appalachian coal seams that run continuously from Alabama break off at a headland at Sidney, Nova Scotia, to reappear a thousand miles northeast on the west coast of Greenland. From there they run under the Greenland ice cap, appearing next on the east coast of Greenland, and then jump across the North Atlantic to reappear in Spitsbergen, 12° from the North Pole. Thence, on a southeast course, this same coal forest jumps deep ocean gaps to reappear in Wales, the Ruhr in Germany, and Czechoslovakia.

While Gargantuan events assaulted the lands of the planet, the only breeding places for a "land animal" that were not fatally altered were down on the coast near where celestial tides sway and eternal waves play. Through the eras, the Teleostei (bony fishes) clung to their fresh-water breeding places. In my mind's eye, I see salmon ancestors finding spots to spawn along coasts when they returned from the ocean each spring—through millennia. Geologic changes are hardly perceptible year by year.

During these epochs the Teleostei had plenty of company. Vertebrate evolution was in high gear. I see amphibians, ancestors of frogs and salamanders, playing around the fresh-water breeding places, and some lizards and water snakes. And during those periods the Teleostei divided into two congregations, when some of those which went to sea to grow up never returned to the ancestral, fresh-water breeding places, and became the haddock and cod, tuna fish and flounders of our time. Probably a gene mutation enabled them to spawn in salt water, while *salmon*, although they also went to sea to grow up with the others, held on to the fresh-water breeding places of the original backbone fish on earth.

In so doing, due to population explosions of many creatures in shore waters—and of predators, which now included big bony fish—crowding the sea upon the continental shelf, our salmon had to seek places for breeding as far upstream and inland as they could reach. The script for the great salmon drama—written in their genes—never permitted eggs to be fertilized in salt water. This can only occur in fresh water instilled with a peculiar chemical quality derived from infusions of algae and the dissolved bodies of fresh-water animalculae, and decaying vegetable matter. The salmon came to have a sort of monopoly on this energy liquor, highly spiced with vitamins, for nourishing strong offspring. The formulae evolved with the biota along with the developing river systems.

For the climax of this saga we turn from those dim, timeless eras to the salmon we behold today arching their backs at the base of a waterfall, leaping, splashing, struggling against fast water.

The great salmon-thronged river systems of Alaska, British

Columbia, and the Northwest, were wrought by the last Ice Age, which still lingers just over the horizon. The cold, clear streams of the Arctic tundra—with its peculiar odor—which tap the snowfields on the mountains, are still in contact with the Ice Age. Torrents that poured out of melting glaciers descended into valleys, left ponds sprinkled far and wide, while they carved sinuous courses among the mountain ranges in their rush to the ocean.

While the Ice Age was at its height, only 20 thousand years ago, massive continental glaciers burdened and sterilized the regions which are now the inland empire of the salmon. Down on the coast, river deltas were everywhere, fresh-water spawning places must have been countless in the inlets and fjords, with ocean right at hand. During the recent Ice Age the salmon must have been fabulously prosperous.

Then, as the shrinking glaciers drew back and retreated up the valleys, the streams running into the sea flowed from sources ever farther inland, and from higher and higher among mountains. Then successive generations of salmon, seeking ever more security, ever more privacy, spawned higher and higher through the centuries—from estuaries in the deltas, to tributaries, to brooks, to ponds in high valleys and far-away tundra, mile after mile.

The powerful muscles, the leaping ability, the "strength of instinct that draws the salmon upriver to the stream where it was born," evolved year-by-year, step-by-step, in pace with its developing river system in the wake of the retreating glaciers.

A CALIFORNIA BEACH SCENE

Consider the stability of that inheritance code on the DNA tape in the physical control of hormone and nerve systems, and of sex organs, as dramatized in the way three kinds of fish—salmon, grunions, and eels—were so strongly imprinted by the fresh water and land conquests of their ancestors that the memory and the habits formed in that long past age have never dimmed. In amazement we see them summoned to the continent,

197

still compelled to exert superfish energies to reproduce! DNA has a prodigious memory.

This is being written just two hundred years after Junipero Serra (Father Juniper), ignoring reports about dangerous, worthless country "up there," led a band of friars from Baja California along the coast to San Diego. Northward from there, along trails no white man had traveled before, he established 21 missions, each one a day's journey from the next, all the way to San Francisco, which he named after St. Francis, the patron saint of his Franciscan order.

Some personal notes by a red-headed man named Juan Colorado, nicknamed "The Flame," have survived to tell us what things were like when Father Juniper was following the coast —all the way on foot because he had "an infection and couldn't sit on a hinny."

> Great flocks of birds covered the bay and Indians put seaweed on their heads, and corkwood under each arm, to float among the birds and quietly to pull under water ducks that pleased them, to drown them, and so gathered ducks as one might pick figs from a tree. . . . At certain tides sandy beaches were hidden under infinite numbers of sardines left by retreating waves. The fish overflowed onto the land as a brimming bucket sheds its burden. We ate until we were completely filled with this delightful fish.

Today, if you drive along the coastal highway which overlooks the beaches in the San Diego area, on nights when the moon is full in June and July, cars are parked wherever a person can climb down to the beach. And as far as the eye can see, bonfires sparkle with family groups ready with frying pans for a rare dinner of delicate fish about 5 inches long—surely the "sardines" of Father Juniper's time.

A stretch of a few miles between Point Conception and just across the border on the coast of Baja California is the only place in the world where these writhing silvery apparitions appear, April through July, from their mysterious sanctum sanctorum in the ocean where they hid in safety for three years. In fish files they bear an apt Roman name, *Leuresthes tenuis*, "slender with smooth garments." They are popularly called "grunions" (grunters), although the only grunting that is heard comes from people scrambling to catch them with bare hands. Nets and traps are forbidden and the bonfire parties are permitted only after June 1. The beach is reserved for grunion trysting in April and May.

They come riding to the beach in the crests of waves on the first or second night after a full moon—thus synchronizing with the highest spring tides. Since date and hour are predicted by the almanac, everybody is ready and watching for silvery glints like a peculiar phosphorescence in the sea water. Someone shouts *"Here they come!"* It is the vanguard of grunions right on schedule at the hour of high tide. The seventh wave after that is fraught with squirming bodies, and then for a time every roller originating in deep water offshore reflects the silvery moonlight with sparkling bodies. A tumbling wave catapults its fish toward the beach, they swim vigorously in the rush of water to attain the farthest reach of the sliding water; their inbred impulse is to be left stranded by the retreating wave.

It looks as jumbled as a game of jackstraws, but the grunions are playing a serious game of reproducing as ordered by their genes.

Each female is escorted by several males, and the little wedding party rides and flaps to the very edge of the pausing wave. Then quickly, while the sand is still wet and soft, the female digs in with her tail by turning and twisting until she is "up to her armpits" (her pectoral fins), and lays a clutch of eggs. While this is happening the males arch over her and as she struggles to release herself, loosening the sand around her body, the males pour on their milt, the eggs are promptly fertilized. All this takes less than half a minute, whereupon females and males together flounce down the glistening sand to catch a ride back to sea on the next retreating wave.

This leaves the fertilized eggs buried deep enough to retain moisture and just beyond the reach of the waves until the next full-moon tide a month later. During that month, the baby grunions make ready to hatch in the sun-warmed sand. When the waves again slide far enough up the beach to wet and loosen the sand above the hatchery, the baby grunions break out and ride the retreating waves back into the ocean. They are only a quarter of an inch long and generally overlooked by people wading around trying to pick up slippery adult fish from the water that rushes back and forth between their legs.

Biologists who are properly preoccupied with laws of nature would disdain the word "miracle" in connection with this ritual. But I invite my readers to think of a better word to describe a performance that perpetuated lives of these fellow travelers through eras because they possessed a particular sense of deep ocean and of a sandy beach, and a sense to feel the cycles of the moon and tides—who knows but what they use the stars also to find their way from mid-ocean to the coast? Something is inborn in them to make them expert in the use of tides, and in joining the rhythm of surf and of waves spreading on the beach. Their every act is timed to the second. So are the stages of their growing. This is not automatic; some may linger in the ocean homeland two years, some three years, until they are mature and have the muscles to emerge and cope with surf, fighting their way up the beach. It is as though somebody in charge of this ritual blew a whistle at the right time.

Biologists say that the commands come from their male and female gonads, which release microscopic chemical particles, hormones, that fire them to reproduce. At this point inherited habits take command and, inspired by genes, they rise from the dark depths to the surface, mount the waves, and ride them to the beach. But naming a particle a hormone, and calling a DNA molecule a gene, does not make the affair less miraculous.

I can't resist one more speculation about this fantastic demonstration of the interplay of life, water, and land.

That grunions never fail to turn up for the beach party punctually is not so singular. So do mackerel and shad; so do migrating birds, and monarch butterflies. Gonad cycles are invariable in many kinds of life. But why do grunions turn up *at that particular place*, and nowhere else in the world? Other tribes of fish are widely distributed by ocean currents. For example, when the American shad come out of the sea to spawn, they use rivers all along the Atlantic coast. But, with the same beaches stretching hundreds of miles north and south, the grunions use only those few miles near the Baja California and the California line. Let us speculate about this.

The grunion sex rite evolved and was imprinted in genes in a very ancient epoch. The mere two hundred years since Father Juniper enjoyed grunion dinners is only an instant in the time schedule of the ocean life of these fish. The base rocks of Baja California and the southern coast ranges of California show that the mountains we see now were uplifted by orogeny about 200 million years ago. This was some 150 million years before the Rocky Mountains and other inland ranges of the Cordillera uplifted. Prior to that, sharks swam in a shallow sea that covered all our Midwest.

The geologic evidence seems to indicate that in the period when Pacific water was lapping a beach near Pittsburgh, Baja California and the southern part of California stood above the sea as a headland marking the western border of the continental block. During the Cenozoic Era, huge events of mountain-lifting and volcanoes on the Pacific Coast caused the withdrawal of the Pacific's transgression, and built the California coastline as we see it. Underlying rock at the head of the Gulf of California and in the Salton Sea area reveals that the headland we have glimpsed in the San Diego area was probably an ancient island which later became connected with Baja California and with the coast northward.

I see grunions in the plumes of waves arriving triumphant on the beach of that isolated island 200 million years ago. They had evolved in a particular "niche" in mid-ocean, and at spawning time the land had a peculiar fascination as the place of origin of their kind. Here was a sandy beach, not a fresh-water stream or a brackish estuary. But they evolved a way of spawning and hatching eggs in wet sand, and a way of getting there and back at the right time, by steering according to the precise phases of the moon.

This succeeded for millennia while the place was an island, during which time the grunions became ineradicably imprinted to reproduce at that spot according to the positions of the moon and stars. After it was connected to the rest of the coast, they did not know the difference. It remained a relatively isolated and undisturbed place right up to our epoch. Today the laws of the Fish Commissioner are needed to give the grunions their margin for survival.

The grunion legend is comparable to that of many extinct and vanishing animals and birds which are relics from bygone times because they could survive in the ocean or on an island surrounded by ocean. In both situations they enjoyed the abundant, never-failing food resources of the ocean, plus relative isolation and protection from predators on islands or other hideouts. On and on their generations rolled. What genetic stability there is under these circumstances!

There were the *great auks* on rocky Funk Island off the east coast of Newfoundland. The last pair was clubbed to death at Eldey Rock off Iceland in 1844. There was the *dodo* on Mauritius Island in the Indian Ocean. It disappeared about 1680 when mariners snatched them for "chicken" dinners. There was the *moa* that lived in New Zealand for 25 million years, until men arrived—I have seen the 12-foot tall skeleton of a moa bird in a museum in Wellington. There was the *elephant bird* of Madagascar, reported by Marco Polo in 1298, but believed by no one until elephant bird eggs were discovered in 1851—a single egg had a volume equal to more than 144 chicken eggs, and weighed some 18 pounds.

The *Komodo dragon* still owns a small island in Malaysia, surrounded by seething surf on coral reefs. W. Douglas Burden, adventurer, somehow managed to get through the reefs in 1926 to photograph the ugly brute from the age of reptiles; with daring, ingenuity, and big hemp ropes, he caught it and presented it to the Bronx Zoo where, deprived of its sustaining surf on coral reefs, it soon died. And of course, there is the famous exhibit a few horizons south of the grunions, the Galápagos Islands with its giant tortoises and a veritable zoo of exciting creatures from a past epoch.

16

The Call of the Deep

This is a spooky adventure in a night in November. The night is black, blacker than most nights. The sunset has been liquidated by a gloomy mist, and then promptly the tangible outdoors—the outlines of the hills against the sky, ghostly lakes, upland meadows—disappears and the dismal droning of an all-night downpour sets in. It is the weather, the season, and the hour for a clique of fishermen in the Catskill Mountains to pursue their hobby.

Around midnight a man is sloshing through an upland meadow, playing the beam of his flashlight on the rills winding down the slope. Suddenly the questing beam spots a 5-foot "black snake" winding through the sopping wet grass. A stranger to the area might well be terrified at nearly stepping on the uncanny thing, but a member of the exclusive eel-watchers fraternity has found what he is looking for.

Eel-watching in a meadow is a peculiar sport. The best site is a gently sloping field among foothills with ponds and brooks nearby. It will have a lush cover of grass and wildflowers, perhaps some scattered bushes, and trees may hide a pond at the top of the meadow. An experienced eel-watcher knows his fishing ground by the lay of the land and he can predict the hour of arrival of eels by the distance from a pond or stream of departure. Traditionally the best fishing is at midnight, and the darker the night, the heavier the downpour, the better.

Runnels from dripping herbage twine through the grassroot jungles following the course of gravity among the contours of the ground. These guide the eels, which are also fluid and sinuous, on their overland trek. They may come from a landlocked pond which turns out to be a dead-end road on the way to the coast, or from a brook that peters out. These dead ends do not deter the eels, which now set out across country to find another stream that has a positive feel of a current which will keep going—perhaps to a lake with an outlet down an ample creek, to a river tributary, to a mainstream, to the *coast!* Eels are informed by their genes to know instinctively that a hillside meadow on a black rainy night is a good overland highway on the way to the ocean.

Get to the ocean they must or else there would be no eels, for they cannot breed in the lakes and streams where people see them and which are to all appearances the native homes of eels.

For one reason, there are no males around in inland water—all the eels seen in lakes and ponds and streams are bachelor girls. For another reason, after some five or ten years of contentment in fresh water and mud, the aging female feels lonely, yearns for male companionship, and at the same time the chemistry of her body fluids is steadily changing so as to transform her from a freshwater fish into a saltwater fish! She must reach the sea as quickly as possible, because after her body fluids have altered to a certain point, she can survive only as an ocean fish.

It might seem that many eels in landlocked lakes, perhaps hundreds of miles from the sea, might never make it to a spawning mate. But this is to underrate the resources of the mating drive.

The most abundant eel areas of North America are lakes and streams of Canada's North Woods and northern New England. Many of these are surrounded by dense spruce forest, and they are surely landlocked, with no visible outlet. But North Woods lakes and ponds are all sweet-water with a gentle flow because this is continental shield country with underground outlets between rock strata, through which eels slide invisibly with the greatest ease.

The city of Halifax, Nova Scotia, stands at the head of a 10-mile indentation of the Atlantic coast. The shore of the deep bay has countless seaweedy coves, tidal marshes, and little bayous among boulders which teem with all sorts of seacoast creatures —larvae of barnacles and starfish, baby crabs and minnows that proliferate on the diatom plants heaved into the cubbyholes on the swell of every tide. Fed by measureless waterways in spruce back-country and on this threshold between continental granite and the North Atlantic, with miles of lush hideouts protected from violent surf, Halifax Bay became the great trysting place for eels on the American coast. Here were the hangouts for the male eels, which never go inland, but grow up in eelgrass tidal marshes and brackish jungles waiting for respective generations of females to return.

Of course this was all arranged centuries before the founding of the city of Halifax—which some day would need a City Waterworks. It was no problem at all for the engineers to build a dam at the outlet of a splendid chain of lakes with sluices that could be thrown open to flush the city sewers.

Come April, there ensued one of the weirdest urban traffic jams on record!

The word traveled to ardent females far inland up the streams from lake to lake—here was a quick shortcut down to the haunts of the boys. No eel need wriggle overland, no one need use a narrow treacherous way under the rocks—just yield the body to the rush and ride the city sluices. The turning up of eels in waterclosets in Halifax (and Dartmouth, Nova Scotia, also) became a *cause célèbre*, about which studies in waterworks engineering journals were published.

Now, if ever, is the time to marvel at the ways of love. Both salmon and eels are of the same order of animal—Teleostei, bony fishes. To reproduce their species, both enact an extravagant drama in which an individual is alternately a freshwater and an ocean fish. Salmon's is a splashing, colorful affair, and their breeding place is clean sand or pretty pebbles shimmering in a clear pool. In contrast, eel's is a dark mystery, in which sombre fish hide in mud, expose themselves in the air in the darkest night, and breed in a black abyss. Moreover, the film is run in reverse!

When the whistle blows for a female eel she quits her mountain retreat as fast as possible and heads for the ocean. She descends the continent and encounters males, so it seems, quite by accident, where they have been waiting down on the coast lazily gorging and sleeping in estuary mud. What happens thereafter to put populations of eels into lakes and ponds and strew them in muddy banks of streams far inland has been a super mystery of biology throughout history.

In the last chapter we considered the saga of salmon. Awesome as it is to find ocean fish climbing mountains in the wake of the melting glaciers of the Ice Age, still salmon have familiar fish forms and standard ways of spawning. Moreover, the evolution of the salmon performance in pace with the melting glaciers of the Ice Age is recognizable because the Ice Age still exists in the twentieth century on high mountains and Arctic tundra and we can see the Ice Age impacts on land and sea.

I have seen, on an esker at Etah, 78° 20′ N, a *sea beach* that is some 300 feet above the present level of the ocean, showing how the land has risen where the weight of the ice was removed in that area. A few miles away I saw creatures about an inch long swimming energetically in a pool of clear, icy fresh water at the base of the ice cap. Specimens were later identified as tadpole shrimps (*Eubranchipoda*). They had left their marine counterparts several hundred feet lower down in the ocean.

Musk ox and reindeer have evolved what it takes to live on exposed tundra—that peculiar, acid, infertile alluvium that forms a thin covering on polar permafrost. Eider ducks breed on the ice-blocked shores of Spitsbergen, drawn to there in the spring from their temperate homes in the British Islands. Most astonishing, the pulling power of sunlight 24 hours a day compels Arctic terns to fly some twenty thousand miles round trip annually between the Antarctic ice cap and the glacier-laden mountains of Spitsbergen.

206

Yes, the beautiful, extravagant salmon and their ways are plausible in terms of our living world, but not so the provoking and inscrutable eel.

THE OUTLANDISH EEL!

When we speculate on the why and wherefore of eels we are entangled with an unreal creature which seems to be left over from an interplay of water and life in an era before the Pleistocene Ice Ages—that is, even before the salmon formed their habits and their muscles to drive upstream, a million years ago. Indeed, the eel lineage branched off from the bony fishes, and they became queer amphibians in the Triassic, *200 million years* ago. The evolutionists who coin name tags for shadowy beings in that epoch call those eel ancestors *Apoda* (no legs); they lived like earthworms by burrowing in wet ground. Eels retain the instinct to burrow in mud.

When the Apoda became frightened and needed to move overland in an emergency, without limbs for locomotion, they developed muscular elongated bodies which could move quite fast by undulating along with belly on the ground—so it is that eels still "swim" on land!

Survival of the species was so well served by Apodas by a combination of hiding in mud and swimming both in water and on land, that this style of bony fish never basically altered while seas invaded lands, mountains were raised up, and continents drifted. Withal, eels never became true amphibians like frogs or salamanders, but they remained Teleostei with a dash of amphibiousness.

The tantalizing enigma of eels is *when and how do they reproduce?* Obviously, sometime, somewhere, eels must spawn and reproduce with great success, there are so many of them.

The subject was fraught with superstitions. Eels prospered magically along with unicorn and phoenix and other fabulous animals in the twelfth century Latin bestiary. Legends arose about them, as with serpents, in Greek mythology.

Around 300 B.C. reproduction of animals by mating was ob-

207

vious and accepted as universal, and the subject of sex became a craze. Aristotle wrote eight books of natural history, entitled *Physics,* in which he probed secrets of life with such perception that today he is called the "Father of Biology." In Aristotle's time there was so much to discover that all life was dazzling. Aristotle, never at a loss for an explanation, taught a fine blend of factual natural history and superstitions, but whatever he declared about living things was "gospel." He did not neglect eels. They were plentiful in the rivers and lakes of the eastern Mediterranean countries in those days and considered delicious to eat, as they are today by many people.

Aristotle, who based his assertions on a rigid examination of facts and physical evidence with which he strove for truth and order, said unequivocally that eels are generated in mud. This keen "Father of Biology," detecting the releasing powers of decay, added that only putrid mud could bring forth eels. This explanation satisfied everybody, including biologists, almost to modern times. In the Middle Ages experimenters dropped horse-hairs into mud, which worked as catalysts to beget eels a few days later. Anyone can see that is logical, since both horsehair and eels are long and slender.

Nineteen centuries after Aristotle, Isaac Walton, the champion of fishermen, deemed that for mud to produce eels it must be seeded by a worm, and he identified the particular worm as one which dwells in the intestines of a black goby, a fish that inhabits muddy ooze.

The long-held belief about the spontaneous generation of eels —obviously the great exception to the sexual reproduction of animals—was attested by the fact that no male and female organs had ever been discovered, although countless eels had been sliced open, and moreover, nobody had ever seen eels mate. Eel research was made in inland lakes and streams—even today people think of them as freshwater fish. Freshwater eels are all females but their female organs do not shape up until after five or six years, when they are en route to the river mouth to meet males in the estuaries. Not until 1874 were the sex organs of both male and female eels discovered and described in biology books. They were found in eels in brackish water on the shore, but those coastal eels were never seen to mate—they couldn't be induced to mate or to fertilize eggs by spawning even in an aquarium.

Microscopes and biological laboratories were laying the foundations for modern biology. It was in connection with investigating the exciting problem of spontaneous generation that Pasteur discovered bacteria, which led to pasteurization of milk. With this leap into a new order of magnitude of life, the mystery of the reproduction of a sizable, conspicuous animal like an eel became a scientific obsession.

And how dynamically they reproduce! In the seventeenth century, eels were a leading industry in Europe. In 1667, Italian fishermen wading into the Arno between two bridges in Pisa in April caught 2,700 pounds of little eels headed upstream in five hours using only hand-held sieves. When the news of this bonanza spread, a smart man waded into the Arno only half a mile from its mouth, and caught 200 pounds of eels in a few minutes at sunrise. They were so thin and tiny it took 1,000 to weigh a pound—but they were meat for a fry. This was the first news of *elvers*, the baby eels.

Evidently eels were born in the ocean. But still nobody saw eels mating or spawning down at the mouth of the Arno or of any other river, nor in shore waters. Eels came down out of the continent and vanished into the ocean in the fall. Elvers came up out of the ocean in countless numbers in the spring. In between, there was a big, and very aggravating, blank.

In 1895, 21 years after the sex organs of eels were discovered, a weird little ghost entered to join the cast of characters of the Saga of the Eel to offer to biologists a key to the solution of the historic mystery.

Fishermen near the coasts of Europe often found certain weird fish in their nets—thin, long ovals like willow leaves, and transparent as glass. The only opaque parts of their bodies were sharp, fierce teeth and a single big black eye. The transparent fleshless little leafy fish, no good for a frying pan or the sardine market, were promptly chucked back into the ocean. For forty years the "willow leaves" were filed away in fish manuals as an odd species of fish labeled *Leptocephalus* (thinhead).

At long last, a couple of scientists came along who had a gift, celebrated by Professor A. D. Moore, called "serendipity" *—it prompts the discovery of things not being looked for; instead of ignoring something, he who has serendipity comes wide-awake and looks into it.

Two Italian ichthyologists fishing in the Straits of Messina stared at a thinhead, which stared back at them with its single big round eye. They shone a light through the transparent willow leaf and saw the heart beat and organs in action. They took a batch of them home and kept them in an aquarium of their laboratory as curiosities. One day they beheld with amazement a thinhead's eye begin to move into another direction. At the same time the organs changed their shapes and assumed new arrangements; the flat transparent body steadily became tubular and dark. For the first time, trained observers were watching the transformation of a thinhead fish into an elver eel—a secret which, until then, had been concealed in estuaries and in tide rips at the mouths of rivers.

So, Leptocephalus was no freak species of fish, it was the *larva of an eel* that had hatched from eel eggs somewhere under the marching whitecaps.

Thereafter, the metamorphosis of eels was pieced together with scientific controls in aquariums where ocean saltiness could be changed to brackish water in pace with the transformation of thinheads, and then a second transformation when they become little elvers with delicious flesh that swim upstream.

But wait! Eggs don't hatch unless they are fertilized. Where in the ocean do those male and female eels go to spawn? Where is "somewhere under the marching whitecaps"? To find the answer, the biggest animal hunt on earth was organized. It called for a genius to organize a scientific plan to search systematically and in detail the whole gale-torn, tossing North Atlantic. It called for a man endowed with inexhaustible patience, zest, and endurance.

Such a man was Johannes Schmidt, a Dane, who in his cele-

---

* A. D. Moore, *Invention, Discovery, and Creativity* (New York: Doubleday & Company, Inc., 1969). "Serendipity" as an intellectual attitude was coined by Horace Walpole in 1754.

brated ships, the *Thor* and the *Dana*, spent 23 years (1904–1927) traversing latitudes, following longitudes, crisscrossing, circling the Atlantic Ocean between the coasts of Europe and America. Schmidt discovered an awe-inspiring solution to the legendary eel puzzle. He reached—almost, but not quite—the total answer. No diver has yet seen or underwater camera photographed the spawning area in the thermocline where the astounding eels have been reproducing, untouched by geologic events, for perhaps 200 million years.

THE STAGE SETTING FOR SCHMIDT'S ODYSSEY

Let us see where the matter stood when Schmidt tackled the problem. Eels were conspicuous and familiar freshwater fish, with no resemblance to ocean fish. They apparently had no fish scales—those are reduced to pinheads buried under the smooth slimy skin. An eel drives forward through the water (and on land) by undulating its long slender body.

In regard to *adaptations* to an environment and way of life, a snaky, slippery body with its wiggly mobility seems perfect today for moving around in mud, winding through wet grass, and above all—this must be what natural selection insisted on— following the curves and convolutions of currents down to the sea, and, on the return trip, weaving upstream through rapids and wiggling in rain-soaked ground to detour waterfalls. An eel, unlike a salmon, is dainty and slender when it comes out of ocean to mount the land. It lacks the brawn to fling itself with a powerful spring action; instead, it corkscrews and shoots upstream like a flexible arrow.

From Aristotle to Schmidt, 2,208 years of human history, it was taken for granted that eels were freshwater creatures. Judging by their distribution in the United States you and I would say so too. Eels are native in Lake Ontario. They are native in lakes and streams of Minnesota, Wisconsin, Illinois, Indiana, and Ohio. They have been stocked in the upper Great Lakes and in some landlocked lakes, seemingly with great success. In these inland waters they flourish and keep growing bigger, some have become giants 6 feet long. It seems that there is no enzyme produced by eel DNA which calls a halt to their growing! It is estimated that some eels stocked in landlocked lakes lived to be 50 years old. But all the eels in those stocked lakes and in the upper Great Lakes die out eventually, for they are but lonely females.

There are no eels in any pond or river west of the Mississippi watershed. It is sheer imagination to say, as did one writer, that elver eels coming upstream in the spring must pass salmon going downstream in rivers of Alaska and our Northwest. It would be more logical to say elver eels wiggling up the streams of our northeast coast could wave to Atlantic salmon coming down.

Mid-continent eels can get up the St. Lawrence River into Lake Ontario, but not beyond Niagara Falls.*

Imagine those eels in the Minnesota lakes making the trip to mid-Atlantic! That's a far, far call of the sea for them to hear! A thousand miles or more to the Gulf of Mexico—across the big Gulf and around the Cape Sable at the tip of the Florida peninsula, and then another thousand miles across the Atlantic horizons to spawn . . . what shall we choose to be astonished about?

In early spring of 1904 a Danish research ship was in the mid-Atlantic on course to the Grand Banks of Newfoundland, the historic fishing area of European countries. Aboard the ship was an imaginative young Dane named Schmidt, who had persuaded his government to let him try a "biological" stunt that might lead Danish fishermen to the best fishing spot.

They were trawling in the plankton just under the surface for newly hatched larvae of commercial fish such as cod, herring, bluefish, and flounders. These are only a fraction of an inch long and translucent—invisible and mostly unknown to fisher-

---

* Do not confuse the true eel (Anguilla) with the wicked lamprey eel, which is a different order of animal. Lampreys look as though they are whistling, with a circular mouth by which they attach to prey fish and suck their blood. They whistled their way above Niagara Falls through the Welland Canal by attaching their vacuum cups to ship hulls, and thus invaded Lake Erie, where they sucked the blood of native Lake Erie fish.

men—buoyant travelers in capricious windblown surface waters. The hope was to come upon swirling clouds of these larvae in the vastness of the ocean—*a single cod* yields about 7 million eggs—and these would show "how the wind was blowing" for good fishing that season.

Schmidt was making a chart of the hauls: the date, latitude and longitude, wind direction, temperature of the water, estimates of the number and kind of larvae. One day he stared into a dripping haul at a *thinhead* 3 inches long! He recalled the Grassi-Calandruccio report made ten years before to the National Academy of Rome about a thinhead caught in the Straits of Messina and how it transformed into an eel—but the only thinheads seen since then were taken close inshore, and here was one in mid-ocean, 1,500 miles from land.

This time it was Schmidt who exercised serendipity. Most people would have shrugged off that "glass willow leaf" as a trifle. Ocean currents carry all sorts of things—after all, Schmidt's assignment was keeping him busy charting the larvae of important fish like cod. But Schmidt pondered that healthy Leptocephalus as it bared its teeth and glared at him with its big black eye. Just for fun, or by habit, he recorded its size (75 mm), the location of the catch to the minute and second, and the depth and direction of the current that was bearing the thinhead on its way when it was caught.

When Schmidt finished his work for that season and returned to his home port in Copenhagen, he persuaded the Danish government to give him a grant, as we would say today, and a ship to fish for thinheads across the Atlantic Ocean. They might lead him to the breeding places of the supernatural eels, which never mate or spawn but which haunt the freshwater lakes and streams of Europe and America.

The Danes, with a long tradition as sea-rovers and fishermen dating back to the Norsemen, held Johannes in high regard. He was resourceful, and valuable to the fishing industries. How long would his research take? Perhaps two or three years? Schmidt made the first entry on his chart of Leptocephalus distribution that day in 1904 when he caught the thinhead 1,500 miles from land. He made the last entry in 1927.

Writing about his exploit afterward, Schmidt said:

> When I accepted the commission from the Danish Government I had only a slight idea of the extraordinary difficulties. The task grew year by year to an unimaginable degree . . . It necessitated cruises of trawling from America to Egypt, from Iceland to the Cape Verde Islands. We found eel larvae all across the Atlantic. . . . Hundreds were taken at great depths . . . during the night about 30 meters (100 feet) below the surface. During the day they were hundreds of meters below. . . .

SCHMIDT'S ODYSSEY

"How long will your research take, Johannes? Two or three years perhaps?" It was only an extension of this energetic young man's oceanographic work. Besides, it would be valuable to find the breeding places of eels—they were an important commercial fish in European markets. So the Danish government backed Schmidt and put at his disposal the sturdy Atlantic trawler *Thor*.

That 3-inch-long "glass willow leaf," which he caught by chance 1,500 miles from land when surveying off the Grand Banks of Newfoundland was the trigger of the whole 23-year project. It whispered to Schmidt that eels must breed in mid-ocean—it was carried in the Gulf Stream—it had come from somewhere. Where? That seemed like an easy assignment for an expert on fish spawning.

He set out on peripheral trawl of the Atlantic from where the Gulf Stream leaves the continental shelf off Cape Hatteras to the area east of the Grand Banks, where he took the southbound fork called the North Atlantic current and headed down to the Azores. The circling of the Atlantic was surprisingly rewarding. Thousands of thinheads were caught and measured—and became pins stuck in charts.

In the following year Schmidt had arranged for trawlers on both sides of the Atlantic to measure and locate thinheads before tossing them back into the sea. The master chart acquired a beautiful pattern of glass eel larvae streaming toward the shores of their respective continents radiating from the same point southeast of Bermuda.

These results explain why the Danish government continued to back Schmidt with ships that kept him heaving among the Atlantic rollers for seventeen years.* When the aging *Thor* was retired, they gave him a sturdy little steamer named the *Dana*, which became equally celebrated.

Schmidt's chart had a peculiar fascination that stirred not only the Atlantic and Mediterranean fishing industries and biologists, but also the general public. It became something other than concentric circles on the broad bosom of the Atlantic—although a true target with a center goal—it was more like a gossamer web of an orb spider which has an eerie beauty with fluid circles intersected by straight "radiating" lines.

On the great eel chart they are not radiating but converging. The straight lines are extrapolated. They represent the courses taken by the big female eels which ride down the rivers from their inland lakes, join the males in the estuaries and vanish in the deep.

As years passed Schmidt's chart evoked the interest of maritime countries surrounding the Atlantic and Mediterranean. Even the Italians joined the chase after they were reconciled to the fact that their eels were not Italian born, but that they streamed through the Straits of Gibraltar at a depth of 400 feet. The masters of fishing smacks, schooners, trawlers, even freight and passenger steamers, offered their services, eager to play a part. They were given special trawling gear and instructed to pause at their convenience, note the precise spot, trawl slowly for half an hour, and report the number and exact size of glass willow leaves in their buckets on deck. Their counts were astonishing—they had never given a thought to those queer flat transparent fish before.

When Schmidt's computer—the pins in the chart—showed him they were closing in on a belt around Bermuda, some

---

* The project covered 23 years—of which Schmidt spent 17 at sea and the rest refining, interpreting, and publishing the chart.

Americans sent out a little schooner yacht, the *Margrette*. She was tragically wrecked in the West Indies by the whiplash of a hurricane.

In his command post in Copenhagen in 1927, Schmidt stuck the last pins in his chart. He had the record of the littlest Leptocephali ever seen—about one-sixth of an inch. Thousands of them had been discovered and hauled up from a few hundred feet beneath the waves. They must have hatched from eggs below the area where these little ones were caught. The area so earnestly sought through the years was finally identified. The dispersion of glass eels all over the Atlantic came to a focus at 25° N/69°W!

The incredible fact that ultimately emerged from this fabulous probe into the mating secrets of eels was that all the eels of Europe from the Arno to the Severn, from 3,000 feet above sea level in the Alps to brooks and ponds in the Hebrides off Scotland—and all the eels of North America, from Nova Scotia to Florida—*meet for spawning in one area*, under the hub of the vast whirlpool in the North Atlantic called the Sargasso Sea. The eerie, dark eel rendezvous is about 1,600 feet below the blue waves.

The mature eels never leave this sunless spawning sanctuary. It is the most currentless and stagnant part of the North Atlantic. It is the center of the huge North Atlantic eddy bounded on the south by the North Equatorial current, on the west by the Gulf Stream which curves northeast and then forks with one branch continuing to the British Islands, even as far as Spitsbergen, and the other branch bending south along the European coast, past the Strait of Gibraltar, and on down to the Canaries and Cape Verde Islands, completing the circle where it meets the North Equatorial current. The still center of this oceanwide whirlpool had been the sanctuary of the weird and ancient eels —even while continents have drifted apart.

So another baffling mystery of water and life, which Aristotle tackled in 300 B.C., was solved by Schmidt 22 centuries later? No—not quite. Nobody has seen those eels on course between their river mouths and the mid-ocean breeding place, or knows what depth they travel. Moreover, as of this writing nobody has ever seen eels spawning, although this will come any day with the perfection of modern techniques for deep sea probing and photographing.

Now it's interesting to regard in the mind's eye the sliding, sinuous motion of eels twixt land and sea. We speak of the "territorial imperative" of animals. Man, rhinoceros, gila monster, wolf, bee, snake, Arctic tern,* and salmon—almost every creature stakes out a claim for a personal home place on the earth's surface and defends it to the death. But not so the eel.

The female resides in the mud of a pond, in either America or Europe, while she matures. But there is no evidence she will defend her mud; eels are not known to put up a fight for anything. Their stay in their freshwater residence averages five years, during which one may slip from pond to pond, a quiet and inoffensive animal never heard to utter a sound.

With a five-year lease for fresh water, 20 percent of the inland eels descend into the ocean annually. We have noted that it takes three years for the European species to reach their continent from their breeding rendezvous in the Sargasso Sea. So the thinheads on route to the Arno, the Severn, the Rhine, any stream in Europe, must pass *three schools* of adult eels going in the opposite direction, I wonder whether they recognize each other as being the same kind of animal.

On the other hand, American thinheads born in the same area reach their coast in a single year—they have only a thousand miles to travel. So—they grow up three times as fast as their European relatives. Their maturing period is paced with their time en route. This is surely dramatic evidence of the way the law of survival operates inexorably according to physical facts.†

Note that I do not say *swimming* time en route to their respective continents. These flat transparent eel larvae ride ocean currents in the manner of the larvae of the ancient plankton—but not passively. They choose the levels in the ocean that have

---

* Arctic tern is the most ferocious on this score. When I invaded a breeding territory to photograph the birds at 79° N in Spitsbergen, I was dive-bombed at perfectly timed intervals by a squadron of sixteen terns, which tore my hat to shreds.

† Although, with our arrogant pollutions, we seem to have no sense of this.

currents going their way. They are feeble wriggling swimmers
with parachute characteristics, and buoyant to be easily borne on
their long journeys while enjoying the nourishment of plankton
soup.

It is impossible to tag thinheads at their birthplace in the deep
vortex of the Sargasso Sea. This could be done only with those
caught when the transparent ⅙ inch eel larvae rise from their
dark hatchery and pirouette by the thousands a hundred feet or
more below the surface. In the deep it would be like tagging
swirling sparks. So, unable to identify an individual arriving on
a coast, nobody knows whether a thinhead goes to the same
river where its parents grew up. Probably not. (This is not
comparable to salmon returning to their birthplace.) Moreover,
freshwater eels, as we have noted, do not have a "territory";
they slither around wherever there is comfortable mud to hide
in. But we do know that the Europeans travel eastward and
the Americans travel westward from their common spawning
deep.*

How do they find currents going their way? How do they
plot a course, shifting from current to current to arrive at the
right continent on schedule? These are searching questions for
which Schmidt had no answers.

When oceanography became a modern science, around 1960,
with undersea research vehicles and atomic-powered submarines, a new map of ocean currents was made, showing the
depths of the ocean flowing *in different directions at different
levels.* Previously, while accepting Schmidt's beautiful and con-
clusive proof of the spawning of eels in the vortex of the Sargasso Sea, baffled oceanographers argued vigorously as to how
American eels could go westward where the Gulf Stream flows
northeastward, and why southern European eels which start
their journey in the Gulf Stream do not end up in the Azores.

Doubtless, thinheads knew all about the thermocline long before today's oceanographers discovered it. (See Chapter 9.)
Moreover, they knew all about the layers of ocean water above
the thermocline—that the Gulf Stream is in the top layer, where

---

* There is evidence that eels seek particular latitudes. There are
probably English eels, Franco-German eels, and Italian eels.

it is propelled in an easterly direction by the planet's spin and by the ceaseless west winds of the northern hemisphere, and that currents in a layer below the Gulf Stream flow in the opposite direction. They selected the currents that would deliver them to their continents probably by keen senses for water pressures at different depths, and for temperatures and salinity at the various levels.

This question about finding the right continent prompts a reverse question. How do adult freshwater, female eels and their male companions down on the coast shape a course to a spawning spot deep in mid-ocean?

This migration surely calls for different navigational skills than those of the famous long-distance migrating birds who have keen eyesight to follow landmarks such as mountain ranges, river systems, and the lines of seacoasts. Those birds which fly over open ocean navigate by the direction of sunlight and by the moon and stars. But eels have no visible landmarks; they see no sun, moon or stars.

There has recently come to light some news about eels that seems to answer this tantalizing question.

In today's oceanographic laboratories, with sophisticated aquariums, scientists are able to find out how fast an eel swims, and they are able to tag individual eels with an isotope of hydrogen called tritium, which is just as good and palatable to drink as H_2O. In this way, they are learning a great deal about how these extraordinary animals navigate in dark and featureless depths of the sea.

Eels are highly sensitive to saltiness, light, and temperature, which guide them in certain directions and to particular depths —just as the temperature gradient in the thermocline guides their glass willow-leaf children to find their way upward to currents which carry them to their continents.

Most sensational of all is the eel's sense of odor. Dr. Arthur D. Hasler, in *Underwater Guideposts*, refers to an experiment in which big eels migrating overland were found to head in the right direction by the "smell of the sea" on a light wind, after they were transported into strange surroundings miles away. This means that the eels we stumbled over at the beginning of this chapter were already sensing the smell of the sea a hundred miles away.

Unparalleled tests have been made by Dr. Harald Teichmann —whose ingenuity and patience are comparable to that of Johannes Schmidt. He arranged a series of dark pipes in a muddy bottom as invitations to burrow, and put a certain odoriferous chemical in one. He trained eels with food rewards to come to the one with this odor. Then he reduced the odor particles (pheromones) until the dilution was equal to that of a single particle in a volume of water "as big as Lake Constance." The trained fish found their way, by smelling a *single odor particle*— ONE pheromone!

This implies that when eels hatch in their sanctuary they are "conditioned" to the precise odor of the breeding place in the deep—and they never forget it. Their response to this odor is indelibly imprinted in their genes.

This is all but incredible, even to molecular biologists. However, it is confirmed by recent electron studies of the nasal equipment of an eel. Unlike ourselves who have quite an area of odor nerves in our nasal chamber (bathed in mucus to dissolve the odor particle), an eel has an extremely tiny nostril, and this leads into an odor cavity so microscopically minute it will accommodate only *a single* odor particle.

According to this evidence, the ocean bestows on thinheads an odor sensitivity which they can pass along to their freshwater female selves which they will use years later as a guide to the spawning place in the black abyss. It is all but supernatural!

"Now here, of all places, is the most charming spot for an ouzel's nest."
So might have said John Muir, champion of the little bird that lives in
waterfalls. (*U.S. Forest Service*)

A view from Storm King Mountain. (*David Plowden*)

H₂O creates spooky architecture when, in freezing temperature, it is lighted by sunlight and pulled by gravity. (*Carl A. Koenig*)

Water in the early spring woods.

"A poet could not but be gay / In such jocund company . . ." Wordsworth

A most hospitable place. This pond is one of thousands that were left by the glacier across our northern states from the Hudson to the Mississippi.

Bartholomew's Cobble—167 acres of native forest, wildflowers, rare ferns. "A little bit left" saved by local people.

17

Twilight on the Delta

The well-watered terrain of our northeast and north-central states—which became the homelands of our largest centers of population—is a glacial delta created by a phenomenon which exerted forces so colossal that this geophysical event of our time seems legendary and unreal. Yet only yesterday, on the geologic timetable (about 20,000 years ago), a lobe of a glacier stood twice as high as the Empire State Building on Manhattan Island. The people who do "carbon dating" of rocks and sediments say that a glacier loomed in Iowa in historic time, around 4,000 B.C.

The ancestors of the Sioux Indians must have seen it about 50 miles east of Sioux City. The huge white dome was a familiar part of the landscape—a landmark for good hunting that was visible from a hundred miles away when it glittered in the sunlight. They saw melt water cascading over a jagged cliff where a section of the ice face had split off and crashed, and water was roaring out of tunnels at the base. Acres of sediments fanned out southward laced by icy streams in a hurry. A hundred yards from the glacier, thickets of blueberry and bearberry bushes abounded in rabbit and fox, and a half mile south groves of birch and poplar were the haunts of deer.

In August 1955 there was an unusual news item in the newspapers. Telegrams summoned archaeologists, paleontologists, and geologists to Turin, Iowa, where a man named Asa Johnson owned a gravel pit. He was bulldozing a hillside of *loess* (a peculiar clay made from rock-flour ground by the movement of a tremendous weight of ice) to get at the gravel that it covered, when a human skull rolled down the slope. He notified the coroner, and together they removed the rest of the skeleton, assuming it was that of a local drunkard who had disappeared—perhaps he had been murdered and buried there.

When Asa's bulldozer tore down the loess and plunged into the gravel underneath, more human bones appeared, including the skeleton of an infant. Asa and the coroner piled them up like spooky jackstraws. A deeper stratum of gravel below these human beings held the bones of wild horses, a camel, a mammoth, and a bison—all native animals of our continent at the time the Big Ice crept so far south. This was a sensational record of events concerning American Indians in the Ice Age that brought archaeologists to crowd Asa Johnson's gravel pit.

The gravel that held the human skeletons was dated 8,000 B.C., and its contact with the loess laid down by a moving glacier is pretty good evidence that ancestors of the Sioux Indians saw the glaciers. Men and wild animals must have thrived on the glacial delta enriched with chlorine, calcium, magnesium, iron, manganese, zinc, copper, aluminum, cobalt, iodine, fluorine, bromine—all vital elements in us and in every mammal body!

The fertility gave priority to green plants—photosynthesis was in high gear on those tundra bogs. Plants use carbon, hydrogen, oxygen, and nitrogen for 90 percent of their *bodies*, but for their enzymes, hormones, and the delicacies of flowering and reproducing—in short, for blooming fertility of forest and meadow, and for tall prairie grass, and ultimately tall corn, and wheat in the sunlight that sways in waves when the wind passes over, and for oaks with acorns, and maples that launch seeds like little airplanes—plants must have potassium, calcium, sodium, magnesium, iron, aluminum, manganese, sulfur, phosphorus, chlorine.

Creeping glaciers played upon by sunlight calling forth the power of rushing water did a magnificent job for life on our continent. This took a lot of time as we measure time by our calendars. The glacier scratched a timetable in rocks of the Connecticut River valley which tells us that the Big Ice retreated from the vicinity of Hartford to the vicinity of St. Johnsbury, Vermont, 190 miles, in 4,100 years. This was the schedule that stocked the rivers with fish, built gay, deciduous wilderness with glorious animals and birds. It does seem like an eternity compared to our mad pace.

However, this power play of sun, water, and ice was controlled by natural forces that took a million years to do a job—the length of the Pleistocene Epoch, which geologists call the Ice Age—to free the vital atoms and molecules locked in granite, to send them swirling dissolved in river systems, to spread them far and wide in sediments across our Midwest, and to be carried in rivers southward far beyond the front of the glacier itself.

The glacial sediments average 50 feet in depth on the prairie. A measurement sounded 300-feet-deep sediments left by the ice at Princeton, Illinois. Beside the Potomac River glacial sediments are 140 feet deep, and even along the James River in Virginia the glacial till is piled 100 feet deep.

It was unknown until around 1850 that the magnificent hospitality of our northeast and north-central states, and all the big river systems east of the Rockies, are the work of glaciers so massive that they depressed the Atlantic coast about 150 feet in the vicinity of Philadelphia, and the coast of Maine was pressed down 200 feet lower than it is today.

Then a young Harvard professor from Switzerland (who later founded the Marine Biological Laboratory at Woods Hole, Massachusetts) was riding a train between Boston and Springfield. His notes of the trip say: "The rapidity of the locomotive is frightful. There is something infernal in the power of steam carrying such heavy masses along with swiftness of lightning—you cry out in dismay."

However, men had tamed the power of steam so that Professor Louis Agassiz could relax enough to look out the train window and contemplate the sandy hills and the strewn boulders. Before he came to Harvard, Agassiz had studied the sediments and boulders on the mountainsides among the glaciers of the Alps. All of a sudden he perceived what made New England landscape special. He scribbled in the notebook on his knee: "All along the road are ancient moraines and polished rocks. No one who has seen these upon the tracks of glaciers could question how these boulders that cover the ground were transported."

Agassiz was so awed by the gifts which the Ice Age had bestowed on America that he regarded it as a divine act—he wrote: "The glacier was God's Big Plow!"

Of course the whole proposition was too big to comprehend and generations of Americans have merely taken for granted the magnificent hospitality of our land, and still do. A million lakes and ponds. Prodigious river systems fed by the copious rainfall from the Atlantic coast to the Great Plains. A million miles of streams with splashing trout and salmon. Green pastures everywhere on hillsides and in valleys. Uncountable trees in illimitable woodlands, with many springs of cool, delicious water —the forests alive with a greater variety of species of birds and animals than any other area of the globe. On the Atlantic coast estuaries teem with ocean larvae, and wingstorms of ducks and geese burst out of quivering marsh grass. So it was before 1950.

Today, viewed from an airplane, the vast glacial delta is clearly discernible and intact. Big bays such as Boston Harbor, Narragansett, New York harbor, Delaware, and Chesapeake are in their places where they were gouged out by the ice. River mouths such as those of the Connecticut, Housatonic, Hudson, and Delaware are blue and heaving with the tides as they have been since the glacier withdrew. There are the magnificent terminal moraines of Cape Cod, Nantucket, Martha's Vineyard, Long Island, and Staten Island, where the Big Ice deposited them.

If you fly westward over New York State you see the slender Finger Lakes with the melic names of the Indian tribes that fished in them and hunted in the woods round about—Otisco, Skaneateles, Owasco, Cayuga, Seneca, Keuka, Canandaigua, Honeoye, Conesus, and Chautauqua. Only "yesterday" their basins were scratched into the granite by the extended arms of the fabulous white octopus.

Continuing westward the Great Lakes, the largest area of fresh water in the world, slide by under the airplane. Draining the great north central basin, they enable shipping to reach the Atlantic Ocean via the St. Lawrence River outlet. All the work of the recent Big Ice!

When the last glacial ice melted away and the water drained from the sediments, gathered in lakes, or went running and splashing toward the sea, plants and animals rushed in. The glacial delta was so vital it literally sprang into life. This is a great paradox—nothing is more lethal than the pressure of glacial ice, especially when it abrades the continental bone. But the summer sunlight that bombarded the ice cliffs and drove them back at last was warming, and it instantly warmed the sediments, and their pools of water. Piles of gravel and single boulders collect heat, and stones that rest on the surface of a glacier will burrow into it vertically by dint of the infrared warmth they transmit.

I have seen both single pebbles and piles of gravel that have bored many feet deep into the brittle, hard-as-steel Greenland ice cap. I have photographed grassy meadows and wildflowers touching the skirts of big Ice Age glaciers at Etah, North Greenland, and in Kongsfjord, Spitsbergen, both places only about 11° from the North Pole. At first this astounded me—I had expected to see scoured and sterile ground. Then it dawned on me that the full spectrum of the sunshine, including the invisible heat-carrying infrared rays absorbed by rocks, was intensely reflected from the glistening ice cliff and from the big chunks of ice all around, and from the splashing, swirling water, which turned the mineral sediments into dynamos of life.

So it was that plants and animals followed close on the heels of the retreating glaciers to lay the foundations for the "magnificent hospitality" in the latitudes of our land where rainfall averages 40 inches a year. That is a lot of water—unfailingly supplied to the glacial delta of our northeast states centuries after the big ice had plowed and gone. One inch of rain falling on one acre of ground adds up to more than 27,000 gallons of water! No wonder our glacial delta has big rivers and lakes, lush pastures and marvelous forests—and springs that gush spontaneously to make a babbling brook or fill a well.

HOW WE HAVE ENJOYED IT!

Until recently the magnificent hospitality was not a measurable quantity. A man could always help himself to more.

The first arrivals came from the northwest. They were the woodland Indians whose ancestors reached America from Asia across the Bering land bridge. It was more than a bridge. Geologists say that when the glaciers of the Ice Age were at their height they locked up so much water evaporated from the sea that the level of the ocean was 300 feet lower, which would make the Bering land bridge about a thousand miles wide. People and animals did not come across that bridge in single file.

Those people did not just pack their belongings and migrate. They had no belongings to pack. They were nomads, always on the move, always looking for better hunting. They followed animal tracks eastward across the bridge—which was then a wide valley carpeted with blueberry and crowberry bushes, and there were millions of rabbits and lemmings, perhaps raccoons, porcupines, wolves, bear, and deer. The ancestors of the woodland Indians must have made their campfires leisurely and they may have taken several generations to cross from Siberia to Alaska. But always they were drawn southward and eastward by the good hunting.

The picture is that of a group of people gathered around a campfire on the bridge, with a stew simmering in a stone pot. The men have long hair and beards. Their limbs and the upper parts of their bodies are bare. A woman holds a baby at her breast; and a boy has a crude bow and arrow in one hand, while the other hand is busy with a thong tied around the neck of a wild dog that is straining at the leash to get at the stew. All the people wear the hides of heavy-furred animals around their waists. In the background the hide of a bear is stretched and fastened to the face of a boulder. It may take a few months for this fresh hide to be properly dried by the sunlight and made ready to wear. But these people have no timetable other than

226

the rotations of the sun and stars and the succession of the seasons.

As men kept following trails that led eastward, they saw taller forests, they followed the tracks of bear, bison, moose, cougar, and wolf. The young hunters took their girls and made their camps farther on, and so it went generation after generation. Their descendants became the woodland Indians who were as native as the bears and beavers.

The land bridge between Asia and America was closed by ocean water only about 10,000 years ago—when the ice in Iowa was melting fast, and the lower Connecticut and the Hudson rivers were in full flood. Then the water from melting glaciers raised the sea level until the ocean flowed through Bering Strait from the Pacific into the Arctic Ocean.

This happened only a short time before our written history began, but the Bering land bridge which delivered the Indians is no myth. It wrote its own history for archaeologists to read about the people and animals that passed that way. Their trails can be traced on each side of Bering Strait—coming out of Mongolia they converge on the "bridge." From the Alaska side they lead southward and eastward into the magnificent hospitality. Just below the surface of the ground are the charcoal remains of their campfires with their stone axes, hammers, flint spearpoints, and the bones of animals which they were hunting and eating.

The magnificent hospitality absorbed and evolved the woodland Indians and they luxuriated in it. Its impact and effect on them was quite different than it was on the people who later came to it from Europe. This is described by Theodore Roosevelt in a book * that he wrote while he was the Police Commissioner of New York City:

> All the wars we waged for the possession of the country between the Alleghenies and the Mississippi were carried on in never ending stretches of gloomy woodland. The underbrush grew dense and rank between the boles of tall

---

* *The Winning of the West* (New York: G. P. Putnam's Sons, 1889).

trees, making a cover so thick that it was in many places impenetrable—so thick that it never gave a chance for human eye to see even as far as bow could carry. No horse could penetrate it save by following the game trails or paths chopped with an axe. . . . Here and there it was broken by a rare hillside glade, or by a meadow in a stream valley, but elsewhere a man might travel for weeks in a perpetual twilight unable to see the sun through the interlacing twigs that formed a dark canopy over his head.

This dense forest was to the Indians a home where they were as much at home as a farmer on his own acres. To their keen eyes, trained for generations to more than a wild beast's watchfulness, the wilderness was an open book . . . with moccasined feet they trod as silently as a cougar . . . they could no more get lost in the trackless wilderness than a civilized man could get lost on a highway.

Some 500 years before Columbus electrified Europe with the discovery of the east coast of "Asia" and staged a circus parade of redskins and red, yellow, and blue parrots in the streets of Seville, Norsemen were enjoying boisterous hospitality on the New England coast. The coastal abundance was fed by the rivers that brought millions upon millions of salmon, trout, and eels down to the sea from inland lakes and streams, and by estuaries where herring gathered to spawn and produce astronomical numbers of eggs—each female produces about 30,000; the eggs are heavier than sea water so they sink and stick to whatever plants or stones or shellfish on the bottom they happen to touch, to hatch when the temperature of the estuarian water is precisely right.

The Norse did not think they were discovering a continent, and they had no curiosity about what lay inland behind the ramparts of the trees that topped the headlands and lined the beaches. They were only island-hopping while exploring for better fishing. This was no new world, only a far-flung extension of their homeland with familiar fogs and surf and rocky headlands crowned with lush meadows and clumps of spruce in valleys.

So the Norse were the first Europeans to enjoy the magnificent hospitality of America. There was good living for them here. Millions of birds squawked and circled above coastal rocks and waded in the estuaries. Their egg-filled nests were dense in the tall grass at the head of every beach. Coastal waters teemed with cod, whales spouted, schools of dolphins broke the water with a hundred turning wheels.

They would have seen thousands of great auks standing erect, 3 feet tall, on offshore rocks and islands around Newfoundland and along the coast of Maine, like sentinels on the ramparts of the continent. The stately birds, straight out of the Ice Age, had black backs and broad white breasts, and big eyes with a large oval white spot under each. The wings were ridiculously little, mere stubs that could never lift that giant bird into the air. But when a great auk saw the water colored by a school of herring, bass, perch, or cod, it dived into the swell of a wave and the little wings whirred like propellers as it swam and darted after the fish as quick and graceful as a seal under water.

When a nor'east storm lifted a tonnage of water and hurled it for three days and nights across the rocks which the great auks owned, the giant birds retreated into their granite fortress under the curves of boulders. They emerged after the storm to lay countless eggs on the bare rocks in the warm sun, and then stand up and gaze again into the ocean. The great auks were unconcerned about gales and crashing surf, but these birds could not defend themselves when men arrived in long boats on a fair day and clubbed them to death as they waddled around completely oblivious to danger.

The great auks were countless—and delicious. For two centuries European ships that visited the Grand Banks, and the Gulf of St. Lawrence, never missed an opportunity to enjoy a fine feast of auks in the fo'c'sle. Jacques Cartier, who brought two small vessels into the St. Lawrence River in 1538, noted in his log: "In less than halfe an houre we filled two boats full of them, so that after we did eat, each ship did powder and salt five barrels of them."

A solitary, final record, after a gap of many years during which no great auks were sighted—a pair of great auks was clubbed to death at Eldey Rock off the coast of Iceland on

June 3, 1844. Whereupon the stately birds which knew no fear vanished from the earth.

That was a solitary, isolated event at a time when the population of the United States was only 2 million, and Florida, Texas, Wisconsin, Minnesota, and all the Far West and Pacific Coast states were not yet admitted to the Union. Yet it serves as a symbol of the beginning of the eroding of the magnificent hospitality at the hands of greedy and mistaken men, to whom the land, the rivers, and the sea appeared to be inexhaustible.

That the mariners were greedy and mistaken would have been farthest from their thoughts. After being cooped up and banged around for weeks in a creaking vessel, they were tired, hungry, and thirsty and craved exercise. Clubbing a few great auks quickly provided relief and good food. After all, those birds were inexhaustible and as permanent as their rocks. During 430 years since Cartier this aura of inexhaustibleness became the American mystique.

It still seems inexhaustible after we write off as past history passenger pigeons, bison herds, tall timber in Maine and Minnesota, Sam Clemens' Mississippi, and William Bartram's bald cypress in southern bayous. To Bartram those superb cypresses were a great gift to men: "The trunks make large and durable pettiaugers and excellent shingles and timber adapted to every purpose in buildings. When the planters fell these mighty trees, they raise a stage around them, to reach above the buttresses; on this stage eight or ten negroes ascend with their axes and fall to work round the trunk that would measure ten to twelve feet in diameter, and fifty feet straight shaft." *

Note that Bartram, who was an ardent lover of the wild, whose sensitive writings about the forests and birds of America inspired the poetry of Wordsworth and Coleridge, felt no dismay at the felling of the giant tree, but rather awe and admiration that it took a lot of men a lot of time to bring down a bald cypress—*they were inexhaustible.*

The final penetration of the impregnable bald cypress swamp forest came in 1942. In that year Roger Tory Peterson slogged

* Helen G. Cruickshank, Ed., *John and William Bartram's America* (New York: Devin-Adair, 1957).

into it "when the axe was already ringing" against the last of the ancient buttresses. He saw a pair of ivory-billed woodpeckers whose notes were like those of an organ in a dim cathedral. "Tooting, staccato, and musical," Peterson said. Alexander Wilson, "Father of American Ornithology," called it a "toy tin trumpet"; Audubon, "a clarinet." Peterson was the last man to hear that tingling music.

There were 140 years between Bartram and Peterson, but it has taken only a single decade, only ten years, to bring the great glacial delta to the brink of final disaster. This delta of our land is a much larger and more impregnable magnificence than bald cypress swamps. Things are suddenly happening appallingly fast. The difference is that between men with hand axes and men with thousands of horsepower.

The American mystique vibrated with fresh vitality when Alaska was admitted to the Union in 1958. "Inaccessible," they called 36 million acres of interior Alaskan virgin forests, watered by the Yukon River—a river *all untouched* that is six times bigger than the Hudson, four and a half times bigger than the Connecticut. Blueprints for a dam promptly appeared on the drawing boards of the Army Engineers in Washington, D. C. The great river, newly born from the Ice Age, cuts through a ridge about 100 miles north of Fairbanks, leaving opposite bluffs facing across almost a mile of valley. This marks the famous Ramparts of the Yukon. Surely a fine place for a dam; why its abutments are already built by nature. The blueprints call for a dam 530 feet high—proudly that would make a fall of water three times higher than Niagara.

The 280-mile long lake behind the dam will not make those inaccessible virgin forests accessible, it will blot them out. In that same fell swoop it will also blot out the homelands of 3.5 *million* fur-bearing animals such as mink, beaver, bear, fox, and moose, according to the estimate of the U. S. Wildlife Service, not to mention dealing a fatal blow to the breeding grounds of uncountable millions of ducks, geese, and salmon. But we have dire necessities for survival—1,500 miles of power lines from a hydroelectric plant on the Yukon can help a lot down our way and perhaps water from Ramparts can steady the level of the

231

Great Lakes. Alaska gives us in our time one more whack at the "inexhaustible" magnificent hospitality.

Adolph Murie, Field Research Biologist, National Park Service, published a book about Alaska's wildlife in 1961.* "Alaska is still a frontier," said Murie. He extolled the "magic" of the primeval landscape where mountain goats and white mountain sheep and moose still wander freely. "Here one may still see the wolverine and bear, and hear the uncanny, mournful music of the northern wolf that have existed in the interior of Alaska, century after century, each following its own way of life. . . . They still exist in the north country."

Why, this is incredible—quite unexpectedly these United States gain a fiftieth state more than twice as big as Texas! Murie is telling us that in the 1960's it had inaccessible interior forests, and tundra plains with herds of wild caribou, and a broad, swift river as big as the Mississippi, where a "rush brook trout in spate" swishes past a bluff, and that 270,000 salmon leap past this same headland annually on the way to spawn in the upper waters of the Yukon.

This jubilee of wildlife is underscored with actual numbers in a report by Paul Brooks published by the Sierra Club in 1965. "This is prime moose country . . . with a carrying capacity of 12,000 moose" not to mention "martens, wolverines, weasels, lynx, muskrat, mink, beaver, otter . . . there are 36,000 lakes and ponds on the Yukon flats alone . . . a nesting habitat which contributes 1,500,000 ducks, 12,500 geese, 10,000 cranes to the flyways of North America annually."

As I write this in the fall of 1969—only eight years after Murie's report, only four years after Paul Brooks'—there are tears in my eyes which make those words, "They still exist in the north country," wobble and become blurred. I have just seen a photograph on the front cover of a magazine which shows caterpillar tractor trails of havoc running across the horizon of the tundra with a large herd of caribou standing there looking at them. It's incongruous, ghoulish. What's going

---

* Adolph Murie, *A Naturalist in Alaska* (New York: Devin-Adair, 1961). Murie lived with the Alaskan animals in their native habitats and writes about them from an ecological standpoint.

232

on? This is answered in the same mail by a front-page story in *The National Observer* which reports that "Tract 57" (on the Arctic Coast of Alaska) was sold to the highest bidder for $72,277,133. It set the audience "cheering, clapping, and whistling."

Thus, before a cheering, delighted audience, begins the next to the last chapter in the story of the magnificent hospitality of a continent.

18

The Great Awakening

Leaving matters to take their course up there on Tract 57, we turn back to the glacial delta, where most of the people of the United States live.

They live here because this is the largest and finest dominion of fertile ground on the globe in the temperate zone, with that 40 inches of rain per year. The rainwater soaks into the soil to be held in the vast water table that underlies all the forests and fields, while an enormous surplus—the run-off—is collected by great river systems in broad valleys where the flow trends from north to south, reflecting the thrust of "God's Big Plow" that carved them out of the bone of the continent.

The giant river systems are well spaced—the Connecticut River, the Hudson, the Ohio, the Mississippi, the Missouri. In between are smaller, more personal rivers with beguiling names —the Androscoggin, Penobscot, Kennebec, Merrimack, Housatonic, Mohawk, Susquehanna, Allegheny and Monongahela, Olentangy, Scioto, Big Walnut, Wabash, the Fox River.

The ponds and lakes scattered around everywhere are the fountainheads of the rivers—sparkling jewels dropped by the glaciers when they retreated over the northern horizons. The Great Lakes are the biggest freshwater reservoirs in the world, with an expanse like the sky, and horizons like oceans. Listen to the names of some of the smaller lakes—Moosehead, Winnipesaukee, Champlain, Oneida, Skaneateles, Cayuga, Seneca, Canandaigua, Wawasee, Winnebago. The rivers and lakes are singing the song of a fabulous well-watered land. Is not its water inexhaustible?

Just because the water was inexhaustible—where fertile sod and mineral ground are deep on hill and dale, where boats can ply lakes and rivers on every hand—the glacial delta spawned bloated cities with their industries and plants, and in the 1950's an appalling population explosion blighted the dominion.

Before that there was rural country between cities, where the characteristic scenery was that of cornfields, pastures, orchards, woodland, and marsh. Just beyond the city line there were truck farms with lush lots of tomatoes, beans, lettuce, pumpkins, and a moiling chicken yard. County roads were macadam and radiated from the county seat with its town hall, store, and church. Then quite suddenly, rows of little houses elongated in all directions. While these tentacles of the cities were seizing millions of acres, farms shrank and became isolated oases.

In the 1960's bulldozers, like fiends out of hell, roared across the countryside with the cruel credentials of eminent domain. They plunged straight ahead through fields and woods; they felled fine trees, filled bogs, obliterated brooks, and tore hideous gashes in mountainsides. In a few years they encrusted 10,000 square miles of the land between cities * with concrete that is much more harsh and sterile than the floor of Death Valley—to meet the crisis of cars created by the population explosion. But are not our fields and woods inexhaustible?

This is an example of the pollution of a vital resource—in this case, just good earth—by shattering and disrupting its contours, patterns of evaporation, and diffusion, and run-off, its bogs and moisture-holding sod under trees, bushes, and grass. But is this not clearly for the public good in the face of an urgent need? And is there not still a good deal of the magnificent hospitality of America left in the 1970's?

I recently drove on the concrete from New York City into southwest Connecticut. After only 50 miles I headed north on

---

* Based on rural mileage figures of the U.S. Bureau of Roads, for the 21 states of our glacial delta.

winding roads, and then a curtain lifted: I was in a wonderland of hillside meadows framed by oaks, maples, poplars, sassafras, their fresh green sweeping up to blue sky, and gorgeous purple loosestrife and cattails in moist hollows. The winding road was bordered by stone walls built with small glacial boulders which farmers had picked up to clear a pasture a century ago. When my car rattled over a bridge, I was startled when kids who were lined up on the railing dived into the creek, vanished in their nakedness quicker than birds. There would be deer, rabbits, squirrels, woodchucks in that area. It would be alive with warblers and robins, and many songbirds in season.

Yes, in 1970 one could still get "off the beaten path," have a picnic under a tree, see some wildflowers, watch some birds. But the fragments of the magnificent hospitality are receding. One must go a hundred miles from a city to find them, and even so, he will catch sight of a lot of little houses through the trees. It is ever more difficult to avoid the concrete that has taken possession of the countryside.

In the 1960's the creeping chaos of defilement of air and water sharply accelerated but was still comparatively inconspicuous. During most of that decade news headlines and outcries on television and radio reflected political and social provocations. Yet people left to the authorities the battles over watersheds and matters that come under the heading of "pollution."

The authorities coped with those matters either locally—as when the lovely Great Barrington in the Berkshires was persuaded to invest in a sewage disposal plant instead of dumping in the little Housatonic River—or, in municipal areas, with measures that would expedite growth and development and lend political lustre as well.

They built bigger, more modern disposal plants—whose stacks contributed lethal fumes to air pollution. They opted for parks—which created litter problems, and dangers of being mugged after dark. They chlorinated drinking water—which made it insipid. They bulldozed junk as "fill" for marshes and estuaries. (Are not high-rise apartments far better than a marsh?) They zoned—and scenes lost to housing and concrete were glorified with a nomenclature, such as "green acres," "riverview," "gardens," and "parkway."

Thus it was through most of the 1960's, with pollution problems inconspicuous among high-rises, and tolerated because of social and political pressures, and with massive water shortages postponed by excess rainfall on the glacial delta.

THE GREAT AWAKENING

In the fall of 1969 came the great awakening over the befouling of our living places. The mess was grown too conspicuous on every hand. Stinking garbage and rubbish smouldering in corners of woods where the town fathers sought to conceal them. Foul waterways—just a few years ago they were creeks where kids could swim and catch fish. Traffic bumper-to-bumper—you are pinned down helpless, with the car in front belching noxious fumes into your nostrils. Yellow smog blurs the modern towers of the city and stings your bronchial tubes. All of a sudden the people heard the rhetoric of the conservationists.

Until the fall of 1969 the conservationists had been regarded as a sect apart from the mainstream of life, with voices crying in the wilderness. That was as long as there was a wilderness to cry in, before ten million people overwhelmed the National Parks in the summer of 1969. The "conservationists" were bird watchers, tree lovers, wildflower lovers, shell collectors—pleasant people, rather interesting to tolerate so long as their tribal rituals didn't interfere with the progress of modern society. Then, quite suddenly, conservation became popular and political.

There are many signs of the awakening. Programs about wildlife and exploration enjoy high ratings on TV. Skin diving is popular. Coral reefs are wonderful. Oceanography competes with the technical sciences as news. Natural history has emerged from its traditional cozy corner and is speaking to us with color photographs and movies about the wonder of life on planet earth. Conservation organizations, which have so long eked out meager budgets, find themselves receiving generous donations.

Their staffs, now on urgent public business, are as busy as stock-brokers in a bull market.

It is the conservationists with their visions who laid the foundations for our survival in the 1970's. They were fighting the battles of salmon rivers, brook trout, glacial ponds in northern Minnesota, redwoods, national seashores—the battles of waterways and flyways—they were involved in the personal affairs of alligators, eagles, and whooping cranes—long before Lake Erie died, Hudson River pollution threatened shad spawning, the big Chesapeake Bay oysters became scarce, and smog smothered megalopolis.

While Theodore Roosevelt was president, just after the turn of the century, he created 38 national refuges for wildlife. Just before his death in 1919 he blew a loud clear bugle call for conservation with these words:

"THOU SHALT NOT PASS. Let this be the slogan of all who wish to see our natural resources preserved for the perpetual use of our people and not destroyed for all time to gratify the greed of the moment." *

T. R. was exhorting conservationists to defend some of America's particular grandeurs—sequoias and ponderosas, grizzlies and wapiti, otters and ouzels—yet what meaning those same words have in the 1970's for *all* America! Today the "natural resources" to defend are rivers, brooks, lakes, bogs, estuaries, seacoast, plankton in the ocean offshore, the loam in the fields, a clump of trees, the water table underneath our feet, and the air above—even a single handsome oak or maple where a surveyor is running a line.

UNPRECEDENTED EVENTS IN THE FALL OF '69

All at once it is no longer a matter of conserving but of *surviving*. People find that they have been puppets in an elemental tragedy, under the sway of hypnotic illusions. Phrases such as "maximum potential for development" and "progress in

---

* James B. Trefethen, *Crusade for Wildlife* (New York: Boone and Crockett Club, 1961).

a modern society" suddenly lost much of their force. When the people woke up, the politicians heard their battle cry of anguish, and some extraordinary things happened.

The Environmental Defense Fund, "established for taking whatever legal action is necessary to protect the environment," filed a restraining order against the U. S. Army Engineers to stop construction of the Cross-Florida Barge Canal. This was the first time that citizens have ever challenged the Army Corps of Engineers to compel them to fully evaluate the social cost of a proposed "improvement."

In the United States Senate, Bill S1818 created an Office of Environmental Quality with power to delay any federal activity potentially dangerous to the environment, pending investigation by "well-trained men, qualified to safeguard our already dented, soiled environment."

In September 1969—at long last—the Secretary of the Interior invoked a new Water Quality Act to prosecute those who discharge massive chemical and sewage wastes into streams and lakes. He began with five big steel and paper concerns plus the city of Toledo, Ohio, and said that polluters of the Passaic River and the Savannah River are next on the list.

Most extraordinary, catching the public by surprise, were the "fall of '69" challenges to huge jetports. *The National Observer* spread this headline across four columns: "ANGRY CONSERVATIONISTS TAKE OFF AFTER BIG NEW AIRPORTS" with reports of stiffening public opposition to jetports at Miami, Los Angeles, Portland, Seattle, Minneapolis, Chicago, and New York. A new kind of pollution is named—"ear pollution" from sonic booms.

Meanwhile, at Miami the vital water fabric of the whole Florida peninsula is threatened. The Department of the Interior produced solemn evidence that a new Miami jetport will inexorably destroy the Big Cypress Swamp, the south Florida ecosystem (the food chain and equilibrium of all living things including man), and the Everglades National Park. This unprecedented pronouncement from on high quashed local promoters who had pooh-poohed the issue by sneering "airport versus alligators."

"Angry conservationists" are nothing new. But never before

have they had such headlines in the national press. Never before have such bills been passed in Congress. Never before have independent, all-powerful bureaus been called to account. Never before has it been good politics to love nature.

When politicians run for office on the issues of sewage and smog, clean beaches and clean water, instead of bromidic "development," "progress," "highways" and "maximum use," it's obvious that the people have had their eyes opened by what they see and smell, they are running scared and taking matters in hand.

They are taking matters in hand in a big way, with "sunrise laws" and with court appeals that have brought stays and postponements of projects in piles of blueprints of the Army Engineers, of industrial expansions disguised as public good, of particular federal and state bureaus poised for irrevocable destruction of some of our natural heritage that seemed to be eternal—until men came armed with eminent domain and a hundred thousand bulldozers.

Those projects could proceed behind people's backs so long as the magnitudes of mutilations were unimagined by the average person. Ramparts Dam on the Yukon River; turning the Grand Canyon of the Colorado into a super reservoir for waterworks and hydropower; carving up Storm King Mountain on the Hudson River for a great power project, where "no river and mountain scenery in the world surpasses the gap through the Highlands in the sunset light"—had been trumpeted as triumphs of technology.

Even more interesting, in connection with the great awakening, is that people are taking matters into their hands in *little ways*—cleaning things up at the local factory, the town dump, the creek where the kids swim. Lowell Thomas, on his *CBS Views the News*, told about a clean-up at Sterling Heights, Michigan:

"Target—an eleven-mile stretch of the muddy, junk-strewn Clinton River. More than three thousand men, women, and children joining in an effort and getting a big assist from local contractors who donated trucks and heavy equipment. Before they were done, the volunteers removing from the river about twelve thousand tons of debris—dead trees, old washing machines, abandoned cars—you name it. . . .

"Clean-up director Al Martin calling it—'a miracle! `Why, we had women up to their necks in water pushing trees and logs to where cranes could get them out of the water.' The newly cleaned river to be used now for boating—its banks for nature trails. All this—and it hasn't cost the taxpayers a dime. A miracle? Sounds like one!'"

So long as pollution is local it can be coped with. It may involve an industrial plant, a town administration, a local water company that can be dealt with by ordinances and citizen action.

A three-sided battle between an industry, a city, and a lake was won by all three in the case of the Solvay Plant, the city of Syracuse, and Lake Onondaga.* The plant made soda ash, used by many industries, from minerals at that location, and the process resulted in an awful odor. So the plant installed chemical waste beds, from which waste water ran clean and pure into the lake. Meanwhile Syracuse had been pouring raw sewage into deep, hospitable Lake Onondaga until it was surfeited and had a foul breath. So the plant offered the use of its chemical waste beds to the city, which soon became clogged and the plant was blamed again. So the company decided to close the factory. Yet this was a major industry and gave good employment. At long last, after four years to raise the money, the city built a modern sewage disposal plant. The insoluble was solved for the people, the industry, and the lake.

Now back to the great awakening of the fall of '69. The mayor of Chicago has talked about "Planned Development" and "realizing full potential" for "the most valuable undeveloped site in the world." That site will be located on platforms not yet built over railroad tracks. An estimated billion dollars (not counting rising costs and inflation) will be spent for this "most effective shot in the arm for the Loop business economy."

The Chicago Open Lands Project replies simply, ". . . and 35,000 people will be stranded in a high-rise ghetto bounded by an expressway, parking lot, the Chicago River, and the Loop."

---

* Peter Briggs, *Water, The Vital Essence* (New York: Harper & Row, Publishers, 1967).

The simple term, *open lands,* has acquired a peculiar fascination. It conjures air to breathe, a pretty lake, glimpses of trees, and it is not abstract. (The worthy word *conservation* has long suffered from being abstract, impersonal.) The Open Space Institute of New York City is aimed at urbanizing communities, encouraging action programs for permanent preservation of open space. They have already saved about 7,500 acres in the New York metropolitan area. A glowing example of the open-space idea protecting a sanctuary is Bartholomew's Cobble, in southwest Massachusetts, a limestone dome that holds a veritable wild botanic garden, featuring ferns, in a bend of the Housatonic River. The Nature Conservancy organization of Washington, D. C., is promoting the sanctuary idea nationally by giving leadership and backing to local organizations. By 1970 the operations of The Nature Conservancy extended from Hawaii to Maine to Florida. "If any natural areas are to be left for the future, they must be set aside today. Once spent, they are gone forever."

Paul Revere rides again, alerting the countryside, calling people to arms. A lot of little people in a lot of little houses are signing petitions—stop filling that bog for a development; stop eyeing that farm for a shopping center; halt that bulldozer aimed at that patch of woods; keep that marsh with its cattails.

A neighborhood meeting is held at the home of Torkel Korling to promote Fox Path Trail. An idea occurs—to plan a long, slim wildlife trail, using an abandoned railroad right-of-way and paths and fence lines. It might be 35 miles long and 10 to 30 feet wide, and be called Fox Path Trail. It would be for hikers, bicyclers, horseback riders—and each locality would line it with bushes, trees, and wildflowers. For this "Volvo Bog must be protected," and "the filling of the slough must be prevented."

All of a sudden, in the fall of '69, folks cared about brooks, bogs, and birds. And when, early in 1970, the neighbors who met at Torkel Korling's house became a vociferous "silent majority," the politicians discovered a new keynote word—*ecology.* Previously, only conservationists and biologists knew what that technical word meant. By 1970 ecology suddenly had the force of a popular slogan. Everybody knew that people can thrive only in relation to other living things in their natural surroundings of air, soil, and water. An extraordinary accident brought this home to the television audience in February 1970.

Early in '69 the wreck of the huge oil tanker *Torrey Canyon* on the Scilly Island reefs off Land's End, England, defiled a hundred miles of coast of southeast England and even the coast of Brittany across the Channel. It surely was a nasty business —but a long way off. A few months later came the oil leak in the sea off Santa Barbara, which fouled that beautiful coast with black tar. With resolute efforts by the oil company, the state, and the people, the beaches were painstakingly cleaned. The treacherous crude oil disasters dramatized pollution of a living place—but they were only news flashes about local accidents.

Torrey Canyon and Santa Barbara were dimming memories in February 1970, when the weird drama was twice repeated. Oil tanker shipwrecks befouled the magnificent coast of Nova Scotia, and the offshore waters that are a treasury teeming with cod and mackerel, and almost at the same time another oil tanker grounded on the west coast of Florida, spreading hideous, cloying black sludge on the sunny beaches where waves of the Gulf spread a luxurious carpet of *Coquinas,* butterfly shells. But Coquinas are for gourmets who collect them for a delicious broth, and for shell collectors searching for butterfly shells with the brightest, most colorful bands. Something else gave a personal meaning to oil spills.

Suddenly people all across the country found themselves relating to wild ducks. People who had not given a hoot about wild ducks before were deeply moved as they watched on television screens tender, loving hands trying to clean the nasty tar off delicate feathers, while the pleading round eyes of the quivering birds looked Americans straight in the face.

Thus, early in 1970, came the call to arms.

President Nixon, on February 16, 1970, sent to Congress a message with a 37-point program for combatting air and water pollution. He requested an appropriation of $4 billion as the federal share of a $10 billion program for building municipal sewage treatment plants.

Two weeks later the governor of New York proposed an appropriation of $930 million (to be spent over a period of a few years) just to clean the water of the Hudson River—to keep pure the magnificent stream whose source is Lake Tear-of-the-Clouds on Mt. Marcy in the Adirondacks, which descends from the mountains through Opalescent Brook.

Also eloquent evidence of the swift and astounding awakening at the start of the new decade—the Justice Department using a 71-year-old federal law to file charges against eleven companies for polluting the rivers and canal in the Chicago area. The charges were brought under the 1899 Rivers and Harbors Act prohibiting the dumping of refuse into navigable waters—specifically in this case, the Chicago River, the Des Plaines River, the Illinois River, and the Sanitary and Ship Canal that serves Chicago. Similar charges were filed against a number of companies in the industrial ports of Newark, New Jersey, and Brooklyn, New York.

A resounding climax to the Great Awakening came on April 22, 1970—it was named EARTH DAY. The brevity of the two syllables brought instant recognition and snatched the now brightly flaming torch from the willing hands of the conservationists and ecologists. The message was as clear as the name of the day—*don't pollute!*

People were fed up. They had been learning and brooding. And they felt so helpless. Suddenly here was something they could do about it. They could parade, they could talk, they could congregate in auditoriums and parks—and *everybody* would pay attention!

Earth Day turned out to be a unique political and social happening. The idea was taken up by some activist groups who had a notion of using it as an issue against the Establishment. Instead, it was a catalyst that united Americans from coast-to-coast, and left the intiators gaping in astonishment.

19

But Thermal Pollution Is Different

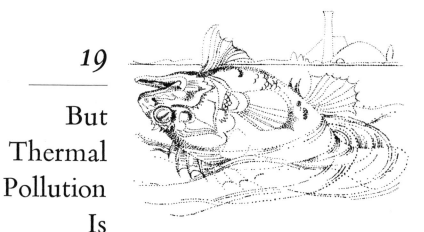

The familiar pollutions are visible and afflict the senses in definite ways. Chimney smoke is black and has a sooty fallout. Fumes from car exhausts smell like noxious gas. Smog makes your sinus ache. Litter and junk are ugly. Garbage dumps reek. Ordinary polluted water looks dirty and is offensive to drink or swim in. But thermal pollution is different.

It is invisible. It has no shape or dimension. It is not a recognizable thing that you can point to over there, up there, or down under there.

Moreover, this impalpable demon has such a disarming name —*thermal pollution*. It means just what it says, "heat pollution." Warm water, discharged into a river or lake from an industrial plant using water for cooling machinery, raises the temperature of that body of water 10° to 35° F. This has been a time-honored resource in all kinds of heat-creating operations. (Look under the hood of your car.)

The term "thermal pollution" originated as a label for an area of research of environmental scientists. It was mentioned in a paper presented at the annual meeting of the American Association for the Advancement of Science, in December 1967. It gives no hint of any menace, and press releases were overlooked in competition with "Atoms for Peace"—which was spurring the proliferation of nuclear energy electric generating, which

245

must get rid of enormous amounts of heat. But after all, what is so bad about warm water?

Warmth quickens life—cold is lethal. The warmth of sunlight makes our planet habitable. Hot water is delightful in a cup of coffee and a tub bath. The triumph of the mammal body is due to its energy generated by the heat of burning carbohydrates. An exquisite thermostat keeps the temperature of your body at 98.6°F, about 30° hotter than comfortable room temperature. In an average-sized adult this life-sustaining heat reaches each one of 30 trillion cells through a network of arteries, veins, and capillaries with a length of 70,000 miles. Five quarts of warm water in the bloodstream make the complete circuit of this aqueduct system about once a minute. In the course of 24 hours 7,200 quarts of warm water course through the body.

Even technical researchers in the atomic energy field felt little or no concern about the hot water by-product of nuclear power generating, as they perfected cool water circulators that would control the hellish heat inside the "pile." The Atomic Energy Commission ignored all public mention of thermal pollution, and dismissed it from consideration until the fall of '69.

This is understandable because the first use of atomic power was in submarines, aircraft carriers, and other seagoing craft, where the fierce heat inside the pile is taken care of by circulating ocean water through the reactor and discharging the hot water back into the sea. Atomic power generating scored a dramatic success in the public mind when the first atomic-powered submarine, *Nautilus*, cruised across the Arctic Ocean under the polar ice, "through" the North Pole, in August 1958.

So the Atomic Energy Commission, riding waves of success from its small-sized seagoing reactors, was ready with the blueprints to respond to the clamor of an electrified populace for an enormous increase of kilowatts for cities. If, in early stages of developing nuclear electric light and power, any questions were raised about disposing of warm water, the AEC, clothed in majesty and mystery, had an easy answer. The atomic plants will be located on a river, on a lake, or beside an ocean estuary.

Studies of the effects of temperature on living things had long been standard in the field of biology—in connection with heart-beats, metabolism, the behavior of larvae, the speed at which a creature runs, and in the control of *blooms*, the excess proliferation of algae and microscopic organisms which discolors lake water. But the term "thermal pollution" was never mentioned in these connections.

Thermal pollution is not a term found in a 734-page volume entitled *Limnology in North America* (University of Wisconsin Press, 1966) with reports by 32 outstanding American biologists whose specialty is water and life. The introduction says that this is "a study of all inland waters . . . of the chemistry and physics of those waters." This conspectus was financially supported by the National Science Foundation and its avowed target was "the dependence of man on water, and the critical problems of supply and quality."

There is no listing of "thermal pollution" in the index of the *Encyclopedia of the Biological Sciences*, 1968 edition. And the phrase does not appear in a book, *Water, The Vital Essence* (New York: Harper and Row, 1967), by an able science writer who calls his book "a report on water pollution."

The publication dates of these works were some fifteen years after the sirens wailed at the experimental reactor at Chalk River in Ontario, evacuating the personnel from a terrible "burn down" when a man at the safety controls pushed buttons 4 and 1 when he should have pushed buttons 4 and 3.* They were nine years after a nuclear power plant at the Windscale Works in England polluted ponds, streams, and fields so that the authorities had to seize all the milk and standing food crops within 400 square miles around the plant.† They were three years after the great Enrico Fermi Atomic Power Plant started operations at Lagoona Beach, near Detroit; ever since, it has been plagued with a series of accidents and has been operating far below its stated capacity due to faulty materials and carelessness.

Before 1966 only nine medium-sized nuclear plants had started

---

* Sheldon Novick, *The Careless Atom* (Boston: Houghton Mifflin Company, 1969).
† Richard Curtis and Elizabeth Hogan, *Perils of the Peaceful Atom* (Garden City, N.Y.: Doubleday & Company, Inc., 1969).

operating—in an experimental way. Extremely complicated and "foolproof" safety systems had been designed to set at rest all misgivings—the specter of the atomic bomb inevitably haunts a nuclear power plant packed with tons of uranium. The atomic scientists, the all-powerful Atomic Energy Commission, and the electric power industry kept accidents that plagued these first nuclear power plants under wraps while they pushed ahead to lay the ghost of the atomic bomb with *plerophoria*—a word that surely describes the temper of the promoters of atomic power electric generating plants. It was coined by Francis Bacon in 1600 and means *overconfidence, subject to a great deal of chance.*

Consider the circumstances. After World War II the Atomic Energy Commission was set up in 1946 to find peaceful uses for nuclear energy. As one writer put it, perhaps we had a guilty conscience and wanted to beat the atom bomb into plowshares. At the same time, in the 1960's, the population explosion brought an overwhelming demand for more electricity, plus more and more air conditioners, refrigerators, dishwashers, TV sets, and for lighting and elevators in towering office buildings and high-rise apartments.

Second, atomic power makes conventional power plants look clumsy and unsightly. Why, a single pound of uranium is equivalent as a fuel to 1,500 tons of coal—250,000 gallons of fuel oil —40 million cubic feet of natural gas. And no black smoke pours out of the stacks of nuclear plants to visibly pollute the air.

Third, atomic energy is riding the tidal wave of Space Age psychology, of which that footprint on the moon in the summer of 1969 is a symbol. "Man can do anything!"

The few nuclear pilot plants—nine of them before 1966—were widely separated. Only local people paid any attention to them. Indeed, the light and power companies appealed to local pride by making visitors welcome at these gleaming, modern-styled power plants which were going to meet the enormous demand for more electricity.

Visitors see a high-roofed, white building 265 feet long, with the proportions of a cathedral. This holds the huge humming turbines. Nearby is a massive dome, 160 feet in diameter, 300 feet high, that holds the uranium *pile* sealed inside a vault with 5-foot-thick concrete walls. Close by the dome the visitor sees a slender, graceful stack, 470 feet high, freshly painted battleship gray with bright red bands. Everything is clean, spic and span. No dingy window panes, no grimy floors, no whisp of smoke coils from the stack that points so high into the clear sky—and one wonders and admires.

Visitors behold the big bright control room with a 50-foot-long instrument panel. The complexity of dials, switches, and flashing red and yellow lights reminds a visitor of the famous Houston control room that we see on the television screen in connection with space exploration. And there is an air of priesthood about the white-coated figures ceaselessly patrolling the dials and flashing lights, and gently touching tiny switches.

The attention of these sentinels must not relax for an instant. There is no chair for them to sit down on; they pace back and forth in silence, only communicating with each other by an occasional word or slight gesture.

Many of these calm custodians of furious atoms have been trained by the United States Navy for service in nuclear submarines, and after enlistment they were welcomed and handsomely rewarded by the atomic power prodigy.

Their attention centers on events in the awesome dungeon of the dome, where the uranium pile is hotter than hell's fires and pressing mightily against the 5-foot-thick cement walls, seething and agonizing to rush out. The uranium fuel pile is installed by long derricks with big claws that hold the parts of this treacherous monster as far as possible from the human workmen; cement is poured to seal the last orifice, and thereafter no man can evermore enter that dome. The atomic chain reactions in the pile (which "could burn a hole through our planet to China") must be handled and their every mood sensed by remote control. Mechanical arms, gears, and levers raise and lower 45 control rods, the mighty heat must be monitored every split second, and the 15-foot-deep level of water in which the uranium pile is submerged must be constantly maintained.

The dynamics of this irritable behemoth, the uranium pile,

are comparable to that of the atom bomb. Uranium pellets about the size of your little finger are stacked end to end in stainless steel 10-foot tubes. Then 204 of these slender tubes, which look as serene as billiard cues, are put together in bundles of 157 tubes each, making a fuel assembly. Finally, these fearful bundles are combined in the *fuel core* that will be interred within the concrete dome. This is the simple arithmetic of the "reactor" entombed in the dome motionless, silent, while it gushes megawatts of electricity into networks of power systems and pours soaring fahrenheits into rivers, lakes, and estuaries.

Handling, transporting, and installing a fuel core are features of the weird drama. With all its housing and accessories the thing weighs 475 tons, and no man must lay hands on it, no railroad flatcar can carry it safely. At the Haddam Neck plant on the Connecticut River they tell with pride the odyssey of their reactor from the place it was constructed near Chattanooga, Tennessee. It was towed by a special enormous barge ponderously down the Tennessee River, down the Ohio, down the Mississippi, through the Gulf of Mexico, the full length of the Inland Waterway, across New York Harbor, the length of Long Island Sound and up the Connecticut River, some 3,500 miles to Haddam Neck! The genius, the human exertion, the passionate idolatry bring to my mind the ancient Egyptians building the pyramids.

SO CLEAN! SO ATTRACTIVE!

The Indian Point plant of Consolidated Edison is approached by a winding road up a hill through a woods with fine old trees. I had visited the place a dozen times before when it was a day's excursion on the old Hudson River Day Line—with woodland paths, wildflowers, birds, picnic areas, and a softball field for office parties. This time, from the top of the rise, the great dome with the "pile," the minaret stack 470 feet tall, and the high-roofed "cathedral" that houses the turbines, flashed into view—in the midst of a battleground—where bulldozers were tearing down trees, upheaving, excavating for two more atomic plants, each three times as big as the present one.

This was the threshold of the Hudson Highlands, where the river is narrow and deep, its flow the main highway between the ocean and upriver breeding grounds of the Hudson's famous shad and striped bass. Directly opposite, across the narrow river, the black hulks of the ghost fleet of World War II freighters added a ghoulish feature to the scene.

The Consolidated Edison made our party welcome with a trip through the "cathedral," where a big rectangular opening at one end framed some trees with green leaves dancing in the wind against a blue sky. A living stained glass window! They gave us a booklet with large colorful pictures of children on merry-go-rounds, and scuba divers looking for delightful under-water life, and some ducks peacefully waiting for sportsmen. The theme: "The new look, the clean look, comes at a time when the electric utility industry is turning its attention to designs to help beautify the nation."

There is some truth in that—but it overlooks, for example, what Dominick J. Pirone, biologist of the Hudson River Fish-ermen's Association, saw in June 1963, the first year of full op-eration of the Indian Point nuclear plant. He says, "On that day we saw 10,000 dead and dying fish under the dock; Con Ed had two trucks hauling dead fish to the dump when the plant was in operation." There was a pile of dead striped bass 12 feet high. A fishing editor visited the scene and estimated the kill as a million fish. But the State Conservation Commissioner's office said that was an exaggeration—he estimated the peak kill at "only 800" striped bass per day.

The power company affirmed that they were installing safe-guards which could surely prevent a repetition of the massacre of the celebrated Hudson River fish. In early 1970 massive fish kills again occurred at Indian Point!

When I visited the Haddam Neck plant on the Connecticut River I drove through some of the loveliest unspoiled country-side of colonial New England, where homesteads with balconies and steep-roofed cupolas still have their stone-block landings on the river which was the main highway in the nineteenth century. Trying to find the nuclear plant, I got lost in winding woods roads—it seemed incongruous—this area was for wildflowers and birds. I came upon a cemetery surrounded by tumbled stone walls overgrown with greenbrier and Virginia creepers—one headstone was dated 1717. Somebody's loved one had slept there for 250 years, covered each October with a fresh red and yellow quilt of sugar maple and poplar leaves.

At a fork of the road by the cemetery two crude handpainted signs were nailed to a tree. One pointed to "Old Rock Landing," the other said "Injin Hollow Road," which led south along the river, and this was the way to Haddam Neck, where the Connecticut Yankee Power Company owned 500 acres. About 80 percent was conserved woodland—the atomic plant cuddles on some 100 acres by the river.

The plant had been in operation for a year and a half before my visit, and in that interval the bulldozer battleground had become a park setting for the clean, exotic "mosque." They had planted locust trees along the river bank, which will have tassels of fragrant white flowers in springtime, and there was a landscaper's dream—a reflecting pool that mirrored a woody hillside.

I credit the power industry with sincerity in that statement "to beautify the nation." It must make them feel better along with the rest of us to do away with smokestacks that billow black smoke, and the railroad tracks and big piles of coal; and to have employees wearing white jackets instead of grimy overalls and smudged hands and faces. They surely do want to "beautify"; besides, it's good public relations.

CAN A NUCLEAR POWER PLANT EXPLODE?

Most people assume—and surely the electric power industry must—that we can trust the authorities and particularly the Atomic Energy Commission with its experts who say there is no danger of an atomic bomb type of explosion. This, I believe, is literally true. The tons of uranium in the concrete dome cannot go *bang*.

In the first place, an atom bomb has a kick-off called a plutonium pile, that is in direct contact with a *solid mass* of uranium. Plutonium is a peculiar element not found in nature. The bomb makers manufacture it in a vacuum chamber with strong magnetic fields called a *cyclotron;* they shoot electrons (particles of negative electricity) with an "atom gun" at uranium, turning it into plutonium. This is packed inside the bomb tight against the mass of uranium. When the bomb is triggered the plutonium sends a jet of radiation throughout the uranium that sets off the all-at-once chain reaction of an atomic explosion. An atomic electric power plant has no such kick-off.

In the second place, the uranium pile is not a solid mass, which is a must for a bomb-sized explosion. The uranium is diluted and the awful energy from the nuclei of its atoms is released at a measured rate. The diluting material is a variety of uranium (that is found aplenty in natural uranium ore) which is much less radioactive, so it damps the fuel pile while also feeding it with energy slowly—like a fertilizer. About 97 percent of the uranium in the fuel pellets is of this kind—a moderator and fertilizer.*

Moreover, the mass is partitioned into small units by stacking the pellets in the slender, 12-foot-long, stainless steel tubes, called *cladding.* These are spaced more than half an inch apart— near enough for an interplay of radiation, separated enough to be controllable. Forty-five control rods made of silver, which absorbs and renders powerless radioactive particles, are distributed among the fuel rods. These are raised or lowered to increase or decrease the atomic heat energy. When raised, they permit more interplay of electricity, raising the temperature hundreds of thousands of degrees as needed. When down, they separate the bundles of fuel rods, reducing their heat energy. Thus the unearthly temperatures in the atomic cauldron deep in the dome are damped by remote control by the sentinels watching the flashing red lights and buzzers in the control room.

Doesn't the vocabulary of atomic energy with bewildering

---

* The scientific term for these two kinds of uranium is *isotopes.* The extremely energetic and dangerous one is U-235. The moderator and fertilizer is U-238.

253

words such as cyclotrons, cladding, plutonium, fertilizing fuel, and uranium with three digits attached, betray that something is going on here which nobody understands? Indeed so! Not one of us visitors to the attractive power plants, no legislator who makes laws to protect the public, not even the members of the Atomic Energy Commission who rule over this giant, can comprehend all its subtleties. Even the scientists carrying on the vital experiments have divided it into disciplines and projects —none can thoroughly know the totality. Their theoretical equations written in the abracadabra of atomic physics can build atomic power plants, but they cannot predict what will happen to the water, or to the air, or to the people living near the plant in the course of time.

That radiation permeates the plant, and perilous particles called *nuclides* * pour out of the stacks, invisibly, is ignored in the rush to respond to public pressure for more and more "juice." A certain amount of this radiation in our environment is "acceptable," but biologists cannot say how much and for how long.

Meanwhile, the leakage of nuclides from the pioneer plants is considered harmless. They have a chamber in which the clothing of visitors is decontaminated. At the Haddam Neck plant they give you a little "dosimeter" to carry in your pocket. When you come out you squint into it to see a needle pointing at a scale that tells how much radiation you have picked up, and whether or not you should go in the decontaminating chamber to be cleansed. There is also a narrow door to pass through with slots in the doorposts that pick up radiations. This is so sensitive that when I held my wristwatch with its radiant dial close to a slot a loud buzzer sounded.

Biological scientists have long been aware of cosmic rays— particles with raw energies that can shatter the nuclei of atoms with which they collide. Where they come from is a profound mystery—they are the Chaos of the Universe. They zip through outer space in streams and bursts, and, thank God, we animalcules are protected from annihilation by this radiation

---

* Radioactive particles released by the uranium fuel.

because the oxygen of air molecules and the hydrogen of water vapor in our atmosphere absorb most of it.

The atomic cauldron of an electric power plant seethes with the same kind of radiation as cosmic rays. The three chief ones are *alpha rays*—these are nuclei of helium gas traveling 5,000 miles a second; *beta rays*—these are electrons torn loose from their atoms, possessed with the intensity of lightning strokes; and *gamma rays*—these are similar to X rays that transit our flesh and bones.

Now man has extracted and stored a mighty lot of uranium from where it had been quietly radiating diffused and muffled in the rocky crust of the earth, and concentrated tons of it in fuel piles planted in the midst of populations of people. The concrete dome is an attempt to imprison this hunk of chaos of the universe—and send some of its cosmic energy through our power lines.

We have noted that the risk of an atomic bomb type of explosion is practically nil. The sleek action of the polished silver control rods effectively dispels this specter, as they smoothly, silently slide up and down.

But nothing is said, either in the Atomic Energy Commission's publicity or during the hospitality hour at the plant, about the possibility of a slow explosion called a *"melt-down"* or a *"burn-down,"* which might generate heat like that of the fires of the sun, melting the "unmeltable" silver rods and cladding, cracking the 5-foot-thick concrete walls.

No law of averages, no mathematical equation, no computer has ever told that this is impossible.

Who dares to say that someday, with hundreds of nuclear power plants, their megaton capacities in full operation, a human error, an imperfect weld, a flaw in a metal, an earthquake, hurricane, an activist's bomb, will not cause inordinate numbers of the 32,028 uranium rods in the fuel assembly to come into alpha-beta-gamma interplay? This would set off a chain reaction entailing a melt-down; alarm bells would clear the plant of people who, as they flee, would be engulfed in tsunamis of lethal nuclides. Unthinkable? Remember the breakdown in November 1965—a complete surprise to the electric light and power people

—which blacked out 80 thousand square miles of northeast United States and Canada?

I have on my desk a list with locations and dates of thirty accidents in connection with nuclear reactors that were caused by human carelessness, structural defects, or faulty materials. Note that a homeowner cannot get any insurance protection against radiation or accidents involving an atomic energy plant —insurance companies calculate risks with shrewd precision.

The dangers that lurk in the dazzling modern-day nuclear plants have been described and decried in detail, and with good authority, elsewhere.* Our interest here is thermal pollution of the water in our lives.

Every atomic energy electric power plant must be located hard by a river, lake, or ocean shore. This is also true of "fossil fuel" plants—they burn coal, natural gas, or petroleum—they too need plenty of water to carry away excess heat. But a nuclear plant throws away *40 percent more heat* than the fossil fuel plants per unit of electricity generated. In this respect it is much less efficient than conventional power plants. Moreover, this greater volume of heat—it is 70 percent waste—which must be expelled by an atomic plant is enormously magnified as the output of electricity generated at a particular site mounts. Atomic plants generate megawatts—millions of watts.

This is a paradox. That tiny pellet you can hold lightly in your hand packs so much energy it is called far more efficient than familiar fuels. Of course! Uranium-235 has some three million times more energy per pound than coal, some two million times more energy per pound than gasoline. But the word "efficient" is in serious trouble the moment you put that uranium to the practical task of boiling water to produce steam to run turbines.

---

* For example: *Perils of the Peaceful Atom,* by Richard Curtis and Elizabeth Hogan (Garden City, N.Y.: Doubleday and Company, Inc., 1969), notable for its documentation. *Science and Survival,* by Barry Commoner (New York: The Viking Press, Inc., 1969). *The Careless Atom,* by Sheldon Novik (Boston: Houghton Mifflin Company, 1969); a review in *Science* (American Association for the Advancement of Science) calls this a "first-rate piece of scientific journalism."

Incidentally, that is all the atomic reactor is doing—boiling water. Neither atoms, nor their alpha-beta-gammas turn into electricity! The turbines of a nuclear power plant are *steam* turbines which generate electricity in the same way as conventional power plants.

All the elaborate, expensive to-do to tame the cosmic giant, all the High Priests and Zealots of Technology, all the rituals which are ushering—rushing would be a better word—nuclear power plants upon the stage of "the 70's," are in hot water. So are Americans who live near Brown's Ferry, Alabama; Lake Keowee, South Carolina; on the Cedar River, Iowa; Turkey Point, Biscayne Bay, Florida; San Clemente, San Diego, California; Humboldt Bay, Eureka, California; on the Deerfield River near Rowe, Massachusetts—where nuclear plants are operating or planned in 1970.

On the glacial delta, at the beginning of the 1970's, atomic water-boiling is thermally polluting from Missouri to Maryland, around the Great Lakes; appearing on the New England coast at Wiscasset, Maine, and Waterford, Connecticut; threatening Lake Champlain, the Finger Lakes, and Long Island Sound. Suddenly nuclear domes are looming beside lakes and streams, and by estuaries and salt marshes on the coast of the province of Father Neptune. These are the areas of the magnificent glacial delta where "there is still a little bit left."

Traditionally, northern New England and New York State have been set aside in people's hearts—and to some extent by law—for camping, fishing, and boating. In Vermont, for example, the fresh air, clean water, the gorgeous fall colors of sugar maples, have attracted fishermen, skiers, and lovers of outdoor life.

In the first half of 1969 it was unthinkable that Lake Champlain—the inland fjord between the Green Mountains and the Adirondacks, stretching from Rouses Point to Fort Ticonderoga—could ever be the site of nuclear plants. In September of that year a spokesman for the Atomic Energy Commission raised the possibility of *two* atomic power installations on opposite shores of Lake Champlain—on Vermont east and Adirondack west.

"Back country folk" have been traditionally served by small electric plants using a local waterfall, or coal cars on a railroad spur. They are being taken by surprise by the megaton atomic plants appearing at their lakes and favorite rivers. Flaunting the banner NUCLEAR NEW ENGLAND, the pioneer Yankee Atomic Energy Company at Rowe, Massachusetts, is combining twenty power companies throughout New England in an atomic grid that poses a threat to any lovely lake or river into which hot water may be dumped.

Local people are horrified to see their neighborhoods "beautified." But their tiny cries are lost in the thunder of the chariots of progress and developments. I have a letter from a friend who attended a meeting in a farmhouse: "In a way, it was pathetic. A fine group of farmers and property owners trying to save their place. They think hard, work hard, donate money for newspaper and radio—looking at them, their will to fight, and their struggle against an unappeasable giant seemed sad. They have no influence, too little money, nothing with which to fight this kind of thing—but they must be active, they must *do* things, that is all."

Those frustrated folk, portrayed by my friend in such human terms, were concerned about the last area of unspoiled Appalachian wildness in Ohio. There, along Clear Creek in the southeastern part of the state, is a priceless heirloom from the Ice Age. Hillsides are carpeted with white and pink *Trillium grandiflorum* every spring, and a gorge holds in moist green silence sacred Rhododendron Hollow, where Indians camped in caves. These people feel—to put it mildly—that the state should not spend millions *to beautify* the place, nor should the Army Engineers *develop* it with a $50 million dam.

A Vermonter tells me that before the resolution of a town committee can be considered, the bulldozers have already excavated—"Ten million dollars are invested before the community can catch up." He added this: "When excavations began for a second atomic power plant at Vernon, Vermont, mothers went out and sat down with their baby carriages in front of the bulldozers."

"And then what happened?" I asked. "They couldn't stay there all night."

The wracking of the hearts of humble people has been expressed by W. Douglas Burden, noted explorer and writer, whose home is on the shore of Lake Champlain: "I would no doubt be regarded by most people as a crank for being so conservation-minded, so deeply aware of the speed at which nature is being overrun by man. So true is it that I would prefer to go back to candles and kerosene lamps than to have the open spaces and green hills of Vermont destroyed for all the benefits of nuclear electricity."

On September 18, 1969, a milestone confrontation of the public and the Atomic Energy Commission took place at the University of Vermont, Burlington, with the governor of the state presiding.

Hitherto the AEC, which had inherited from the World War II bomb-makers sole power over "atomic energy for peace," were dashing ahead full tilt—electric power capacity must be doubled, nay tripled, in the 1970's. Military laboratories and installations—such as the Hanford plutonium plant on the Columbia River in the state of Washington, out of the public eye, with "top secret" labels—had perfected the shining control rods, the delicate combination of U-235 and U-238, the domes with their concrete walls—so let's go!

The Atomic Energy Commission, with full authority to overrule local ordinances and state laws, was devoted to research and action for progress and development—their mood was to brush off the problems of radioactivity and thermal pollution. These were invisible, and they take time to do any harm, if they do at all. The need for more kilowatts was urgent.

The strategy of the AEC was to hold public hearings in which AEC staff members heard college professors and mayors protest. When the routine hearing was held to protest a nuclear plant at Calvert Cliffs, Maryland, $2 million had already been spent by the Baltimore Gas and Electric Company on a beautiful parkway leading to the site. Until the historic Burlington meeting, the board members of the AEC held aloof from the public. Then, no longer able to turn a deaf ear to the insistent buzzing of fears and questions which had increasing scientific authority, the whole Atomic Energy Commission with their staff, 39 strong, swooped down on Burlington in U.S. Air Force planes. They confronted 1,500 critical scientists, the public, and the press.

At Burlington, and again at Minneapolis, Minnesota, on October 10, 1969, the consequences of radioactivity and thermal pollution from nuclear power plants could be predicted with a high degree of accuracy. At these meetings, face-to-face for the first time with the AEC, reports of top scientific authorities —using facts, not theories—were presented.*

Among these facts were: the degrees of temperature of hot water which is thrown away into a lake, river, or estuary by a nuclear plant of a certain size; the seasonal temperatures of lakes and rivers and how the lives of fish, and minute organisms of the food chain, are precisely attuned to that cycle of temperature; the volume of water that runs through a watershed down to the sea each year—and at any point on a river or in a lake how much discharge of hot water will raise the temperature, say 20° or 30° and higher. These facts were presented in tables with an accuracy like that of insurance company mortality tables. A species of salmon dies when the water is above 72°F. Salmon and trout cannot migrate in water warmer than 68°F. Eelgrass in estuaries and salt marshes stops growing at 68°F.

Eelgrass, algae, and seaweedy hideouts are the trysting places of eels, the feeding places of wading and migrating birds, and spawning hideouts of oysters, clams, snails, and barnacles—they

---

* *Thermal Pollution, Its Sources, Control, and Cost*, by Dr. Dean E. Abrahamson, University of Minnesota, Minneapolis—with a superb bibliography. A team of Cornell University scientists, headed by Professor Alfred E. Eipper, detailed the irreparable damage to life in Cayuga Lake, threatened by an atomic plant 12 miles north of Ithaca. Dominick J. Pirone, Hudson River Fishermen's Association, is the foremost champion of fish in the rivers of America. His main target is Consolidated Edison's Indian-Point-on-the-Hudson plant—in August 1969 a total of three huge atomic plants at that one spot had been authorized. The work of greatest scope is that of the Atmospheric Sciences Research Center, State University of New York, Albany Campus. The director, Dr. Vincent J. Schaefer, has assembled 22 top authorities from universities all over the United States and from Europe and India. Their own research, plus coordinating data from other research centers, is unique in studies of radionuclides in the air and thermal pollution of water.

teem with life, they are the dynamo of the great food chain of all life on earth.

Pause for a moment to note what Dudley Lunt, sensitive author of *Thousand Acre Marsh*, says about a wild salt marsh: "The peculiar smell of the marsh in the spring is a curious compound of the luxuriance of the new growth mingling with the rich decomposing of that of the last and all the yesteryears, and it is the first breath of salt air off the ocean. The breeze brings you still more—for your ears a chorus of the strangest calls, clucks, chuckles, cackles, peeps, whistles, quacks, and trills you ever heard, for the marsh teems with bird life. And for your eyes there is further feasting—an expanse of green eelgrass weaving gently in the breeze."

Gilbert Klingel, businessman in Baltimore, the only confirmed eelgrass sitter I have ever heard about, writes in his book *The Bay:* *

> I spent several hours sitting on the bottom in eelgrass after dark. Blades of grass were lighted by phosphorescence as the gentle current brushed small animals against their edges. In the background glowing streaks marked the courses of a multitude of minnows, as they dashed about after the small crustaceans. I could feel their feathery bodies bumping into my bare arms and clinging to the hairs of my legs. Some small jellyfish drifted slowly by and flared into greenish brilliance as they touched some object or were excited by some event important only to jellyfish.

Among remaining marshes on the Atlantic Coast are the 90,000-acre Dorchester marshes and the 11,500-acre Somerset marshes on the eastern shore of Chesapeake Bay. The Dorchester Marshes, with their 11,000-acre Blackwater National Wildlife Refuge, make an outstanding winter feeding ground for waterfowl, and attract the greatest concentration of waterfowl on the Atlantic flyway. The Somerset Marshes, which form a complex of coastal salt marshes, are also outstanding as a winter feeding ground.

* Dudley C. Lunt, *Thousand Acre Marsh* (New York: The Macmillan Company, 1959).

Both of these far-reaching marsh areas produce large quantities of the microorganisms upon which fish, oysters, clams, and crabs depend. They have been referred to as "powerful biological engines." *

At the Battle of Burlington, the sachems of atomic energy faltered. Critical questions put to them by the scientists were not clearly answered. *Nucleonic News*—atomic science's own mouthpiece—reported: "AEC did not do so well—and realizes it must do better. Its people did not field criticism cleanly nor effectively." Burlington betrayed the consequences of nuclear power are not comprehended even by its own practitioners.

For instance, Consolidated Edison, the giant undertaker of nuclear electric power, reassures its stockholders that the hot water from Indian-Point-on-the-Hudson-River is spread on the surface of the river, where it cools without mixing with the bottom water where larvae and small and delicate organisms at the base of the food chain live. This makes sense—hot water floats. But the fact is that a wind phenomenon called *Langmuir streaks* forms parallel streaks (10 feet apart in lakes and rivers, 160 feet apart in the ocean) where the surface water sinks straight down 32 feet to the plankton zone. This weird water wheel carries the surface water down at the rate of 3 inches per second, and the streaks move horizontally across the surface at the rate of 6 miles per hour.†

The promoters of atomic power and the alarmed scientists and angry public promptly chose up sides at Burlington. The AEC and industry had harsh words for the critics, whom they regarded as a "conspiracy," and called them "conservationists" and "survivalists"—with a sneer. The scientists who presented the biological data were called *"ecoschizophrenics"*—ecologists who are nitwits. Some of the opposition to the industry led off with battle cries of frenetic speculation: "Radioactivity is any-

---

* *American Forests*, August 1970.
† Wind causes Langmuir streaks but the cause of the sinking mechanism is unknown. Dr. Vincent Schaefer's group at the State University of New York at Albany is the authority for this phenomenon and its effect on thermal pollution.

where from one million to one billion times more toxic than any chemical poison." "Thermal pollution will wipe out our major aquatic ecosystems within our lifetime." "At the present rate of heat discharge into American rivers, some of those rivers will reach their boiling point by 1980 and evaporate by 2010." Maybe so!

However, the unprecedented Burlington confrontation cleared the air. The AEC learned that all persons questioning nuclear power are not irresponsible and uninformed. On the other hand the AEC received plaudits for at long last meeting the problem head-on with the public. Glenn T. Seaborg, AEC chairman, whose aloofness on the throne has given him an ominous image, appeared on TV to say that he was allying himself with those who want "to keep Vermont green and its air and water clean. . . . What is important is that we do everything possible for the greatest good for our people with a minimum of impact on the environment." So the AEC may not be irretrievable, deep-dyed demons after all!

THERMAL POLLUTION—MALIGNANCY OF LIVING WATERS

So the *efficiency* of that pellet of uranium, no bigger than your little finger, is mocked by the awkward dilemma of getting rid of hot water. All the exquisite arts of our space age cannot compel the pellet to yield its heat energy in measured amounts —as do coal and oil burners.

The electric turbines can use only about 30 percent of the heat produced by the seething pile in the reactor. The excess heat—70 percent of that produced—is useless to the electricity generating and must be got rid of fast. So a great volume of cold water must be sucked into the maw of the reactor—and then disgorged as hot water that has to be gotten off the premises as quickly as possible.

An appalling figure will tell you the magnitude of this hot water problem. The nuclear electric power plants under construction or on the drawing boards in 1970 will require 200 billion gallons of water per day for cooling. That figure is about *half* of all the natural run-off of the United States (except Hawaii and Alaska). This poses the threat that nuclear power plants clustered on principal rivers and lakes will have to channel the whole flow of those rivers and lake outlets through their reactors in low water seasons and during spells of drought.

This was not thought of by fishermen I questioned in the fall of 1969 on the Connecticut River near the Haddam Neck plant. One, a retired college professor, said that his fisherman's thermometer jumped 16° when he was in the broad river opposite the plant. The other said he was seeing only small minnows and other useless tiny fish around his dock "instead of shad and stripers these days." Haddam Neck had been operating for only two years. Perhaps in a few more years, after the new upriver reactors at Vernon, Vermont, have thrown away their heat into the Connecticut, whatever remains for fishermen will be served boiled.

So what are the alternatives?

Many different ideas are being tried, in some of which the AEC is taking an interest. It's surprisingly hard to get rid of the hot water.

You can't solve this problem by building plants on the coast. We have noted that coves, estuaries, and salt marshes are the nurseries of ocean life, particularly our coastal fisheries. How big is the ocean? Indeed, its life is widespread. I have seen the sandy beach on Fire Island, south shore of Long Island, literally loaded with seaweeds, snails, gooseneck barnacles, sea squirts, flotsam and jetsam from the tropical West Indies after a strong three-day, southeast blow.

A report from Thor Heyerdahl after his voyage across the Atlantic on his papyrus boat, *Ra*, says: "Mid-ocean was visibly polluted, even hundreds of miles from land." Heyerdahl said that on five occasions they ran into patches of pollution matter "so thick that the crew hesitated to dip their toothbrushes in the water." A member of the crew of the five destroyers on the "search and rescue" mission for the atomic submarine *Scorpion* confirms Heyerdahl's comments. On a course between Norfolk, Virginia, and the Azores he saw so much mid-ocean pollution that "if the *Scorpion* did go down on this route all sign of her debris would undoubtedly have been lost amid the garbage we sighted."

Another expedient is termed *cooling towers*. These are massive towers, some 500 feet high, an expensive item added to the cost of the turbine and electric generating installations. One of the best designs cools the hot water like a giant automobile; this type is referred to as a closed-cycle cooling system. Another design pumps the hot water to the top of the tower to let it evaporate by tumbling down a series of steps. But this raises a cloud of vapor that may become fog, dangerous to airports and highway traffic, and deposits too much precipitation in one place, which can ruin vegetation in that vicinity.

Cooling lakes sound like a simple solution. Why not just spread out the water in an open reservoir, or create a lovely artificial lake, in which the warm water would cool off to the temperature of the air before going back into the lake or river? The monster is too big to handle in this way. Each 1,000-megawatt power plant now being built would require a lake surface of 2,000 acres—a lake a mile wide and 3 miles long. Where's the land?

Serious efforts are being made to spread the excess hot water upon arid ground and use it to quicken the ripening of crops. But this calls for expensive piping systems, and besides, farmers do not need this kind of help the year round. The warm water must be paced with the crop season and periods of low rainfall. At other times farms would be drowned.

Wherever you turn—atomic energy from those trim, clean plants brings serious repercussions. When man tinkers with the power in the nuclei of atoms, nature has cruel backlashes in store.

WHAT TO DO?

First, admit that the nuclear program is running away with us. Put the brakes on. Stall for time to learn more about it. Meanwhile turn to coal, natural gas, and oil. (Consolidated Edison announces as I write this that they are planning more conventional coal power generators.) Great strides have been made in controlling smoke from stacks. Sulfur dioxide is being removed from burning coal and oil by various economical processes. The by-products can be sold, and it is claimed that the entire cost of the process is offset.

A remarkable new process called MHD (magneto-hydrodynamics) is well along in its development. It eliminates thermal pollution from coal-fired plants and "holds promise of fuel cost savings that will aggregate in the billions." This revolutionary plan is based on directly converting certain gases into electricity with the aid of a revolving magnet. Its efficiency (saving waste heat) will exceed even the best coal- or oil-fired plants, and far exceed the nuclear power plants—as high as 60 percent compared to about 30 percent for the atomic plants.* It's not ready (in 1970) to take over, but pilot plants are operating.

Finally—what is needed is plenty of traditional American enterprise and ingenuity.

A recent issue of *New Scientist* (British) expressed some interesting philosophy which I summarize as follows: The issue is whether human beings are going to retain control of the atomic technologies they have achieved or go under to them. There is a real risk that our man-made environment could enslave us as completely as the natural environment enslaved primitive man. Then he was surrounded by forces he did not understand, and he lived in fear. It could happen again.

I wonder.

---

* *MHD for Central Station Power Generation: A Plan of Action,* Office of Science and Technology (June, 1969).

INDEX

Abelson, Philip H., 4
Actin, 82
Adult, 111
Agassiz, Louis, 223–24
Agulhas Current, 10
Air bubbles, 40
Alaska, 231–33
Algae, 9
blue-green, 53, 67, 75–76, 88–91, 106
Alligators, 171
Alpha rays, 255
Aluminum, 5, 7
Amebas, 78, 91
Amino acids, 52–53
Ammonia, 7
liquid, 76
Amphibians, 55
Anaerbes, 58–59
Andrews, Roy Chapman, 175
Angle of alignment of water molecules, 30–32
Anguilla, 212*n*
Animal evolution onto land, 166–82
insects and, 176–82
Latimera, 169–70
Lingula, 168–69
plants and, 166–67
return to the sea and, 172–76
Annelida, 65*n*, 165
Antennae, 104
Ape, brain of, 92–95
Aphids, 180
Apoda, 207
Archimedes, 43
Arctic tern, 145–46, 206
Argonaut, 159–62
Aristophanes, 77
Aristotle, 151, 208
Arthropoda, 65*n*, 87
Aschelminthes, 65*n*
Atlantic Abysmal Plain, 71
Atmosphere, primordial, 7
changes in, 9–10, 60, 81
Atomic bonding of the water molecule, 27–32
structure of, 33–37

Atomic Energy Commission (AEC), 246–48, 259–63
Atomic power, thermal pollution and, 245–66
Atoms, 28
Australopithecus, 92–93
Auxins, 191*n*

Bacteria, 67
sexuality of, 128–29
Bald cypress swamp, 230–31
Balloons, 13
Barnacles, 63
Bartram, William, 230
Bay, The (Klingel), 261
Bering Strait, 226–27
Berrill, Norman J., 132–33, 139
Beta rays, 255
Big Cypress Swamp, 239
Bilateral symmetry, 153, 158
Biological clocks, 193–94
Biotic World and Man (Milne and Milne), 148
Birch, 180
Birds, 55, 65*n*
Black, Joseph, 12
Blastophage psenes, 181
Blood, octopus, 159*n*
Blue-green algae, 53, 67, 75–76, 88–91, 106
Blue whale, 174
Boerhaave, Hermann, 12
Boiling point of water, 6, 40
Bonding, atomic, water molecule and, 27–32
structure of, 33–37
Bony fish, 172–75, 205
Brain
of apes, 92–95
of cuttlefish, 104
evolution of human, 95–97
human and fish, 173–74
Brightness, diatom gametes and, 131
Brittle star, 117
Bromine, 63
Brooding of seahorse, 150

INDEX

Brooks, Paul, 232
Brown seaweeds, 64, 133–35
Buddenbrock, Baron Wolfgang von, 110
Budding, 85, 132, 144
Bullen, Frank, 162
Buoyancy, 46
Burden, W. Douglas, 171, 202, 259
"Burn-down," 255–56
Butterfly, 112

Calcium, 5, 39–40, 63, 78, 131
Capillarity, 44–45
Carbohydrates, 39
Carbon, 5, 7, 38, 40
Carbon dioxide, 3–4, 38–39
    reproduction of hydras and, 145–46
Carbonic acid, 38
Carotene, 88–89
Carson, Rachel, 64
Cartier, Jacques, 229
Caterpillars, 112, 180
Cavendish, Henry, 13–17
Cell and Psyche (Sinnott), 105–106
Cell precursors, 53
Cellulose, 85–86
Center of equilibrium, 34
Centipedes, 62, 65n
Cephalopods, 152–54
Chambered nautilus, 155–56
Chameleons, 171
Changes in the primordial atmosphere, 9–10, 60, 81
Chemical energy, 89
Chicago Open Lands Project, 241–42
Chlorine, 5, 38, 63
Chlorophyll, 46, 56, 58–60, 88–89
Chordata, 65–66, 87
Chordatella, 130
Chromatophores, 164
Chromosomes, 126–28, 141
Cilia, 112
Ciona (sea squirt), 57, 85–86
Cladding, 253
Clam, 117, 152
Climbing perch, 174
Clouds, 22–23
Cobalt, 63–64
Cocci bacteria, 53, 75–76
Cochlea, 104–105
Cockroaches, 176, 178–79
Cod, 139
Colorado, Juan, 198
Comb jelly, 117

Conger eels, 154–55
Conjugation, 129
Continental shelf, 71–72
Contracting muscles, 82
Convection currents, 124
Cooling lakes, 265
Cooling towers, 265
Copper, 38
Core, earth's, 7
Coriolis force, 49–50
Corn plants, 139
Cortex, 95
Courtney-Latimer, Miss, 169–70
Crabs, 63
    horseshoe, 170–71
Crustacea, 122–23, 169
Cuttlebone, 153, 157–59
Cuttlefish, 104, 156–58
Cyclotron, 253

Dalton, John, 17–18
Darwin, Charles, 79–80, 119
Daughter colonies, 114
Denizens of the Deep (Bullen), 162
Desalination, 42
Deuterium, 4
Diamonds, 42
Diatoms, 69–71, 112
    sexuality of, 130–31
Differentiation, 144
Dipole moment, 38–39
Discovery of the H_2O molecule, 12–18
Displacement, 43
Dissolving power of the water molecule, 37–40
DNA, 56, 141–43
    of bacteria, 128–29
    imprinting on nerve and hormone systems of salmon, grunions and eels, 197–202
Dodo, 202
Dogfish, 173–74
Dolphins, 94–95, 172
Dougherty, Ellsworth C., 129
Dragonflies, 176–77
Dry ice, 3–4
Dust cloud, origin of the earth and, 5–7
Dynamics of moving water, 44–45

Eardrums, 103, 105
Ears, 104–105
Earth Day, 244
Earthworms, 65n

Eels, 154–55, 197–98, 203–20
  reproduction of, 207–11
  spawning of, 211–20
Eggs, 138–41
Electricity, 13–14
Electrolysis, 27
Electrons, 28
Electrostatics, 28, 31
Element composition
  of mammals, 5n
  of the ocean, 62–64
  of plants, 5
Elementa Chemiae (Boerhaave), 12
Elephant bird, 202
Elvers, 209
Embryo, human, epigenesis of, 119–20,
  143
Energies of water, energies of orga-
  nisms and, 45–50
Energy cycle, 51–52
Entropy, 25n
Environmental Defense Fund, 239
Enzymes, 56
Epigenesis, 119–20, 143
ESP (extra sensory perception), 105–
  106
Etna, Mt., 21
Eubranchipoda, 206
Evans, Howard E., 179
Evaporation, 27
Everglades National Park, 239
Evolution, reasons for, 81
Evolution onto land, 166–82
  insects and, 176–82
  Latimera, 169–70
  Lingula, 168–69
  plants and, 166–67
  return to the sea and, 172–76
Extraterrestrial life, 3, 74–75

Fabre, Jean Henry, 108–109
Fabric of the water molecule, 33–50
  dissolving power of, 37–40
  polywater, 45–48
  skin of water, 40–45
Ferns, 65
Figs, 181
Fingertips, 103
Fish, 55
Fission, 129, 131–32
Flagella, 112
Flagellates, 80
Floor of the ocean, 71–72
Florida, 239

Flowers, 65n
Fluorine, 63
Fossae, 103
Franklin, Benjamin, 13
Freezing point of water, 6
Frogs, 112
Fruit flies, 145
Fruits, 65
Fucus, 134
Fungi, 65n

Galápagos tortoise, 171
Galvani, Luigi, 13–14
Gametes, 130–31, 138–43
  histone and, 141–43
Gamma rays, 255
Garfish, 173–74
Garstang, Walter, 121–23, 127
Gene codes, changes in, 124
Genero, 16
Genes, 141–43
Giant squid, 162–65
"Gill books," 170
Glacial delta of the U.S., 221–33
Globigerina, 72, 78–79
Gold, 42, 64
Gonads, 133
Goode, Ruth, 141–42
Granite, 7
Great auks, 202, 229–30
Great water cycle, 10, 36–37
Growth and Form (Thompson), 155–
  56
Grubworms, 112
Grunions, 46, 197–202
Gulf Stream, 10, 49
Gypsy moths, 108
Gyrinidae, 43

H₂O Molecule, see Water molecule
Haddam Neck power plant, 251–52,
  264
Halemaumau, 21
Halobates, 44
Hardy, Alister, 68, 84–85, 118, 121
Hasler, Arthur D., 186–90, 193–94,
  219–20
Hearing of fish, 102–103
Hectocotylus, 151, 160–61, 165
Helium, 4
Hells Canyon, 35
Hermaphrodites, 132–33
Heuvelmans (author), 152
Heyerdahl, Thor, 264–65

INDEX

Hindenburg disaster, 19–21
Hippocampus Kudo, 149–50
Histone, 141–43
Holmes, Oliver Wendell, 156
Honeybees, 107–108
Hormones, 191*n*
Horseshoe crabs, 170–71
Houseflies, 107, 112, 179
Hudor, 16
Human brain
  evolution of, 95–97
  fish and, 173–74
Human reproduction, brown seaweed
  and, 134–35
Humboldt Current, 10
Hydra, 143–46
Hydrogen
  discovery of water molecule
    and, 12–16
  origin of water and, 4–11
  oxidation of, 20–21
  properties of, 25–26
  See also Water molecule
Hydrogen sulfide, 26*n*
Hypothalamus, 93

Ice Ages, 172–73, 223
Indian Point power plant, 250–51, 262
Indians, migrations of, 226
Insects, 64, 65*n*, 96, 176–82
  Nauplius larvae and, 123
Iodine, 63, 64
Ion, 28
Iron, 5, 7, 38, 62
Isosceles triangle, 35
Isotonic, 185

Japan Current, 10
Jellyfish, 83–84, 105, 117
Jetports, 239
Johnson, Asa, 222
Journey into Summer (Teale), 178
Jupiter, 8

Keller, Helen, 107
Kelp, 63, 134
Kingdom of the Octopus (Lane),
  153–54
Klingel, Gilbert, 261
Komodo dragon, 171, 202
Korling, Torkel, 242
Krill, 175

Lamp shell, 117

Lamprey eel, 212*n*
Land, animal evolution onto, 166–82
  insects, 176–82
  Latimeria, 169–70
  Lingula, 168–69
  plants and, 166–67
  return to the sea of, 176–82
Lane, Frank, 153–54, 160
Langmuir streaks, 262
Larvae, 111, 116–24, 125–28
Latimeria, 169–70
Lavoisier, Antoine, 15–16
Leafhoppers, 179, 180
Leptocephalus, 209–10
Leuresthes tenuis, 198–99
Lichens, 65*n*, 75–76
Life, role of water in creation of,
  51–60
  proliferation of, 58–60
  protein and, 55–58
  See also Oceans
Light, 88–89, 106
Limestone, 7, 39–40
Lingula, 63, 168–69
Lipmann, 51–52
Liverworts, 65*n*
Locomotion of plankton, 67–68
Loomis, W. F., 145
Lungfish, 174
Lunt, Dudley, 261

MacDiarmid, Professor, 3
MacGinity, Professor, 140
MacMillan, Donald B., 149
Magnesium, 5, 7, 38, 62
Magnolias, 180
Mammals, 55, 65*n*
Manganese, 5, 42, 131
Marine Institute (Monaco), 116–17
Mars, 3–4, 8, 75
Mayflies, 176, 177–78
Meganyctiphanes, 68
Meiosis, 133–34, 143
"Melt-down," 255–56
Melville, Herman, 153
Mendel, Gregor, 121*n*
Mercury, 8
Metamorphosis of sea animals, 111–24
Metazoa, 81
Methane, 7, 26*n*
MHD (magneto-hydro-dynamic), 266
Microspheres, 53
Miller, Benjamin F., 141–42
Milne, Lorus J., and Margery., 148

Milmosa, 106
Miner, Roy Waldo, 153–54
Mineral composition of the oceans,
    62–64
Mitochondrion, 56, 62
Moa, 202
Moby Dick (Melville), 153
Mollusca, 65*n*, 152
Moncrieff, R. W., 83
Moon rocks, 74
Moore, A. D., 210
Moray eels, 154–55
Morgan, T. H., 145
Moss animal, 117
Mosses, 65*n*
Motions, dynamics of water and, 44–45
Mucous, 110
Mudskippers, 174
Murie, Adolph, 232
Musca domestica, 107
Muscles, contracting, 82
Mutations, 92
Myosin, 82

Narwhal, 148–49
Natural selection, 119
Nauplius, 122–23
Nematodes, 65*n*
Neptune, 8
New System of Chemical Philosophy
    (Dalton), 18
Nickel, 7
Nitrogen, 5, 38
Nixon, Richard M., 243
Noctiluca, 80–81
Nomeus grovoni, 84–85
Notochord, 85
Notonecta, 44
Nuclear power plants, thermal pollu-
    tion and, 245–66
Nucleus, 28
Nuclides, 254
Nymphs, 177

Oceans, 4–11, 61–73
    diatoms, 68–71
    floor of, 71–72
    gathering of waters and, 9–11
    origin of the earth and, 4–6
    origin of waters and, 6–9
    plankton and, 66–68
    return to by land animals, 172–76
    rotation of the earth and currents of,
        49–50
    sex in, 125–35

Ocelli, 104
Octopus, 117, 150–65
    argonaut, 158–62
    chambered nautilus and, 155–56
    cuttlefish and, 156–58
    giant squid, 162–65
    highest form of organic sea life, 152–
        55
    stories of, 150–52
Odor, 92, 109–10
    odor memory of eels, 219–20
    odor memory of salmon, 188–91
    See also Senses; Smell
Office of Environmental Quality, 239
Olamic union, 29
Oozes, 72
Oparin, 51, 53–55
Opposable thumb, 94
Orchids, 181
Organisms, energies of, and energies of
    water, 45–50
Origin of the earth, 4–6
Origin of life, role of water in, 51–60,
    61–73
    diatoms, 68–71
    plankton, 66–68
    proliferation of life and, 58–60
    protein and, 55–58
Origin of the oceans, 4–11
    gathering of waters and, 9–11
    origin of earth and, 4–6
    origin of water and, 6–9
Origin of water, 6–9
Oscillatoria, 88
Osculum, 132
Overcrowding, diatom gametes and,
    131
Oxidation, 20–21
Oxygen, 12
    origin of water and, 4–11
    properties of, 26–27
    See also Water molecule
Oysters, 117, 132–33, 139–40, 152
Ozone, 66
Ozone zone, 26–27

Pandorina, 113, 128
Paramecium, 55, 106
Peptide chains, 53
Peter Pan hormone, 141–42
Peterson, Roger Tory, 230–31
Phaeophyta, 134
Pheromone, 109–10, 189–90
Phosphene, 26*n*

INDEX

Phosphorescence, 79–80
Phosphorus, 5, 7, 38, 62, 65
Photosynthesis, 58–60, 88
Phytohormones, 191n
Pigments, 88
Pistils, 181
Planets, 8
Plankton, 61–62, 64–65, 66–68, 76–81
Plant lice, 180
Plants, 65
  element composition of, 5
  as first living organisms on land,
    166–67
  insects and, 180–82
Pleodorina, 128
Pliny, 151–52
Pluto, 8
Plutonium, 253
Pollen, 140
Pollution, 235–44
  nuclear power, thermal pollution
    and, 245–66
Polywater, 46–48
Portuguese man-of-war, 83–85
Potassium, 5, 7, 38
Praying mantises, 160n
Precipitation, 22–23
Pre-geologic phase, 10
Priestly, John, 12n
Properties of hydrogen, 26–27
Properties of oxygen, 26–27
Protein, 45–46
  myosin and actin, 82
  origin of life and, 52–53, 55–58
Protons, 28
Protoplasm, polywater and, 48
Protozoans, 112–13
Pseudopodium, 78
Pyramids, 33
Pythagoras, 33–34

Radiolaria, 72, 78–79
Ramparts Dam (Alaska), 231–33, 240
Regeneration, 144, 154
Rematocysts, 143n
Reproduction, 77
  brown seaweed and human, 133–35
  of eels, 207–11
  of the octopus, 160–61
  survival and, 118–20
  of Volvox, 113–14
  See also Sex in the ocean
Reptiles, 55, 65n
Retinas, 103

Rhodopsin, 88n
Riley, C. V., 181n
Rivers and Harbors Act (1899), 244
Roc, 149
Rocks, hydrogen and, 7
"Roof head," 177
Roosevelt, Theodore, 227–28, 238
Root hairs, 65
Rotation of the earth, ocean currents
  and, 49–50
Roundworms, 65n
Rozier, Pilatre de, 13, 15

Salmon, 139, 145
  spawning and life of, 183–98
Sand dollar, 117
Sap, 167
Sargasso Sea, 216
Sargasso weed, 68–69
Saturn, 4, 8
Schmid, Johannes, 210–16
Sea, *see* Oceans
Sea anemone, 117
Sea cucumber, 117
Sea hare, 140
Sea horse, 149–50
Sea lily, 117
Sea squirt, 57, 85–86
Sea urchin, 117
Seaborg, Glenn T., 263
Seaweed, 71
  brown, 64, 133–35
Seeds, 65
Senses, 77, 87, 103
  human brain and, 95–97
  of multicellular organisms, 91–92
  sensitivity, 103–106
  sharpening of individual, 106–108
  sight in blue-green algae, 88–90
  touch in blue-green algae, 90–91
Sequoia trees, 139
Serpent star, 117
Sex, smell and 108–10
"Sex Gas of the Hydra" (Loomis),
  145
Sex in the ocean, 125–46
  brown seaweed and human repro-
    duction, 133–35
  DNA and histone, 141–43
  gametes, 138–41
  hydra, 143–46
  primordial, 128–33
Shad, 200
Sharks, 102–103

Shell, 28
Shellfish, 152
   primitive, 62
Shrimp, 132–33
Sight, 77, 87
   blue-green algae and 88–90
   See also Senses
Silicon, 7, 38, 63
   Radiolaria and, 78–79
Silver, 42, 63
Silverfish, 176
Sinnott, Edmund W., 105–106
Skin of water, 40–45
Skinner, Brian J., 74
Slime molds, 65n
Slippery water, 48
Slugs, 65n
Smell, 77, 87–88, 92
   sex and, 108–10
   See also Senses
Snails, 63, 65n, 117, 152, 160
Snakes, water, 171
Sockeye, 183–84
Sodium, 5, 7, 38
Solutions, 38
Sound, 77, 87–88
   hearing of fish and, 102–103
   sound waves, ocean and, 98–102
   See also Senses
Spawning of eels, 211–20
Spawning of salmon, 183–98
   DNA imprinting on nerve and hor-
      mone systems of, 197–98
   odor memory and, 188–91
   in open seas, 191–94
Species, 66
Sperm
   cells, 138–41
   of sponges, 132
Spiders, 65n, 160n, 179
Spiraling drain motion, 48–50
Spittle bugs, 180–81
Sponge, 82–83
   budding of, 132
Squid, giant, 162–65
Stability of world water supply, 22–
   23
Stamens, 181
Starch, 60
Starfish, 117
Structure, atomic, of the water mole-
   cule, 33–37
Sugar, 60
Sulphur, 5, 7, 38, 63, 65

Sun, 4
Surface tension, 40, 42–43
Survival and reproduction, 118–20,
   136–38

Tadpole, 112
Taste, 77, 87–88
   See also Senses
Taste buds, 103
Teale, Edwin Way, 178
Tegeticula alba, 181–82
Teichmann, Harald, 220
Teleostei, 186–87, 196, 205
Temperature range of water, 6
Territorial imperative of eels, 217
Tetrahedron, 34
Thermal pollution, nuclear power
   plants and, 245–66
Thermocline, 100, 169
Thermodynamics, laws of, 25n
Thompson, D'Arcy, 155–56, 157
Thousand Acre Marsh (Lunt), 261
Thrips, 180
Thyroid gland, 64
Thysanura, 176
Torrey Canyon (tanker), 242
Touch, 77, 87–88, 91–92
   blue-green algae and, 90–91
   See also Senses
Transmogrification, 64
Trees, 139–40, 166
   plankton and, 64–65
Tritium oceanography, 100
Turtles, 171

Underwater Guideposts (Hasler), 219–
   20
Upper Amazon River system, 10
Uranus, 8

Valley of Ten Thousand Smokes
   (Alaska), 21–22
Vanadium, 57, 64
Venus, 4, 8
Venus's-girdle, 117
Vesuvius, Mt., 21
Vitamin B$_{12}$, 131
Volcanoes, 9, 21–22
Volvox, 113–14
Voyage of the Beagle (Darwin), 79–
   80

Wald, George, 3
Walton, Isaac, 185, 208

INDEX

Wasps, 181
Water ice, 4
Water molecule (H₂O),
    atomic structure of, 27–37
    discovery of, 12–18
    oxidation of hydrogen and, 20–21
    spiraling drain motion and, 48–50
    See also Fabric of the water molecule; Hydrogen; Oxygen
Water Quality Act (1969), 239
Water snakes, 171
Watt, James, 14–15

Wavelengths, 88–89, 106
Wetness, 38
Whales, 172, 174–75
White bass, 186–87
Worms, 62, 117

X chromosomes, 126

Y chromosomes, 126
Yucca moths, 181–82
Yukon River dam, 231–32, 240